Engineering Properties of Agricultural Produce

Engineering Properties of Agricultural Produce

Edited by

Suresh Chandra

Associate Professor, Department of Agricultural Engineering
Officer in Charge, Department of Food Process Engineering
College of Post Harvest Technology and Food Processing
Sardar Vallabhbhai Patel University of Agriculture & Technology
Meerut-250110 (UP)

Samsher

Professor, Department of Agricultural Engineering
Dean, College of Post Harvest Technology & Food Processing
Dean, College of Agriculture
Head, Department of Food Processing Technology
Sardar Vallabhbhai Patel University of Agriculture & Technology
Meerut-250110 (UP)

Suneel Kumar Goyal

Assistant Professor
Farm Engineering, (Agricultural Engg.)
Institute of Agricultural Sciences, Banaras Hindu University
RGSC, Barkachha, Mirzapur-231 001 (UP)

Durvesh Kumari

CDTRI (PCDF), Meerut (UP)

NEW INDIA PUBLISHING AGENCY

New Delhi – 110 034

First published 2021
by CRC Press
2 Park Square, Milton Park, Abingdon, Oxon, OX14 4RN

and by CRC Press
6000 Broken Sound Parkway NW, Suite 300, Boca Raton, FL 33487-2742

CRC Press is an imprint of Informa UK Limited

British Library Cataloguing-in-Publication Data
A catalogue record for this book is available from the British Library

Library of Congress Cataloging-in-Publication Data
A catalog record has been requested

ISBN: 978-0-367-63500-8 (hbk)

Dedicated
to
Food Engineers

Foreword

Agriculture sector is not only vital for food, nutritional security but also remains the principal source of livelihood for more than 58% of the Indian population where majority is vegetarian in its food habit. India's population has been growing at an annual rate of 1.8%, and is expected to touch the mark of 1.3 billion by the year 2020. The green revolution has transformed our state from food scarcity to food sustainability. Agriculture and the food industry are closely connected industries. The food industry is usually divided into food manufacturing and the eating-out industry. Even though food manufacturing is part of the manufacturing industry and the eating-out industry is a service industry, the two industries are relatively closely related as they both require agricultural products as raw ingredients. Increasing urbanization, consciousness on health and nutrition and changing lifestyle are changing the consumption habits of India. The number of working women, single students/ professionals and nuclear families is increasing creating a demand for processed Ready-to-eat foods. Growth of organized retail, which makes the processed food readily available, is also driving growth of Food Processing. India also produces a variety of temperate to tropical fruits, vegetables and other food products. Processing of food products plays an important role in the conservation and effective utilization of fruits and vegetables. India's strong agricultural base, variety of climatic zones and accelerating economic growth holds significant potential for food processing industry that provides a strong link between agriculture and consumers. Agricultural produce and by-products are perishable in nature in varying degree and their perishability gets exploited on the market floor compelling distress sales orchestrated by factors of demand and supply, intervention of the faces of marketing in the absence of matching post-harvest technology (PHT) and agro-processing infrastructure. Agricultural Engineering inputs are also needed to assure remunerative prices to the growers and a share in the value addition to the growers through on farm PHT and value addition to their produce and by-products in order to strengthen their livelihood base landholdings are decreasing for their socio-economic sustenance and assure minimum standards of living.

The engineering selection and design of food processes and equipment requires knowledge of the properties of food materials. These properties are of great importance in the simulation and design of food processes and in the computer-aided process engineering. Their influence is even greater in problems of conceptual design, in which a wrong estimation of a property can lead to an infeasible design plan. Not only the knowledge of properties aids in engineering design and control but also gives information about the product quality, its acceptability by the consumer of different groups and its behavior post production, during storage, during consumption and post consumption. The importance of economical production of agricultural materials, especially crops and animal products serving as base materials for foodstuffs and of their technological processing (mechanical operations, storage, handling etc.) is ever-increasing. During technological processes agricultural materials may be exposed to various mechanical, thermal, electrical. optical and acoustical (e.g. ultrasonic) effects. To ensure optimal design of such processes, the Interactions between biological materials and the physical effects acting on them, as well as the general laws governing the same, must be known.

It gives me immense pleasure to know that a book entitled "Engineering Properties of Agricultural Produce" is prepared by Dr. Suresh Chandra, Prof. Samsher and Dr. Suneel Kumar Goyal, Durvesh Kumari and I am sure, this book will be definitely an excellent source of information to the readers.

(Rakesh Bhatnagar)
Vice Chancellor
Banaras Hindu University, Varanasi, (UP)

Foreword

Engineering properties are the properties which are useful and necessary in the design and operation of equipment employed in the field of agricultural processing. They are also useful for design and development of other farm/agricultural machineries. As we know that various unit operations such as cleaning, grading, drying, dehydration, storage, milling, handling and transportation, thermal processing of biological materials are among the important operations in agricultural processing. In these operations while handling of grains and other commodities the properties which play an important role are physical, mechanical, frictional, rheological, aerodynamic, hydrodynamic, electrical and optical properties of the agricultural produces. Basic information on these properties are of great importance and helpful for the engineering students, engineers, food scientists, extension scientists, extension workers and of course processors towards efficient process and equipment development.

I am happy to know that an attempt has been made to describe some of the engineering properties usually encountered in post production handling of agricultural produces in the form of an edited book entitled "Engineering Properties of Agricultural Produce" prepared by Dr. Suresh Chandra, Prof. Samsher, Dr. Suneel Kumar Goyal and Durvesh Kumari.

I wish all success for this publication.

(Ramesh Chand)
Director
Institute of Agricultural Science
Banaras Hindu University, Varanasi, (UP)

Foreword

Food and Agriculture is the largest sector of the national economy. The food system comprises of production of food, its post harvest handling, distribution and consumption. To retain its quality and nutritive value, it is essential to ensure the integrity and safety of the food through the food chain. Foods are materials, raw, processed, or formulated, that are consumed orally by humans or animals for growth, health, satisfaction, pleasure, and satisfying social needs. Generally, there is no limitation on the amount of food that may be consumed (as there is for a drug in the form of dosage). The term food processing involves the application of principles and practices of engineering, microbiology and food chemistry. Food engineering means application of engineering of principles and practices on agricultural produce of farm and animal origin.

In order to design the machines and equipments used in crop or seed plantation, harvesting, transportation, storage, processing and oil extraction from oilseeds, there is a need to know engineering properties of agricultural produces. The knowledge of physical properties of food materials is of importance to plant breeders, engineers, machine manufacturers, food scientists, processors, and consumers. The data on physical properties are used in designing relevant machines and equipment for harvesting, handling, transportation, separating, aeration, sizing, storing, packing and the other processing. The data have also been used for assessing the product quality. The knowledge of engineering (physical and mechanical) properties constitutes important and essential data in the design of machines, storage structures and processes. The value of this basic information is not only important to engineers but also to food scientists, processors, and other scientists who may exploit these properties and find new uses.

I am pleased to write the foreword of this book which contains scientific information about the engineering properties of agricultural produces. There is immense need of such kind of book in the field of Agricultural Engineering and Food Processing. This book presents various information from user point of view, where the various engineering properties such as size, shape, volume, density, porosity etc. of food material are discussed very well. I am happy with the effort made by Dr. Suresh Chandra, Prof. Samsher, Dr. Suneel Kumar Goyal and Durvesh Kumari.

I am assured that their book will be useful for students, teachers and all others who are interested in this field.

Prof. Gaya Prasad
Vice Chancellor
Sardar Vallabhbhai Patel University of Agriculture & Technology
Meerut (UP)

Preface

India is an agricultural country and various types of agricultural commodities are produced in large quantities. To handle such quantities during harvest, post harvest, processing and transportation, various types of equipments are required. To design particular equipment or determining the behavior of the product for its handling, physical properties such as size, shape, surface area, volume, density, porosity are very important. Various types of cleaning, grading and separation equipment are designed on the basis of physical properties of seeds such as size, shape, specific gravity etc. The densities of grain determine the size of screening surface. The shape of the product is an important parameter which affects conveying characteristic of solid materials by air or water. The shape is also considered in calculation of various cooling and heating loads of food materials.

We tried to write an edited book to provide a fundamental understanding of engineering properties of agricultural produce. In this book, the knowledge of engineering properties are combined with engineering knowledge. Each chapter in the book will be helpful for the students to understand the relationship between engineering properties of raw, semi-finished and processed food to obtain products with desired shelf-life and quality. This book discusses basic definitions, principles of engineering properties and their measurement methods with research findings. It will be helpful to the students for their self-study and to gain information how to analyze experimental data to generate practical information. It will also be helpful for students who deal with engineering properties in their research. Methods to measure these properties are also explained in details.

We would also like to thank our contributors and teachers/colleagues for their inspiration and timely submission of manuscripts. Last but not least, we would like to thank our family members for their continuous support during writing of this book.

Finally, we are making a plea to those who make use of this book to supply us with information or points of view that differ with those expressed in. We know that there will be errors, for which we alone are responsible, and we will appreciate the opportunity to correct those. Feedback from readers of this book are invited to feel free to write us their valuable opinion, critical comments and suggestions will certainly help us to improve its quality in future.

Editors

Contents

List of Contributors

1. **Abhinay Shashank,** Centre of Food Science and Technology, Institute of Agricultural Sciences, Banaras Hindu University, Varanasi, Uttar Pradesh
2. **Amit Kumar,** Deptt of Agril Engg, Sri Karan Narendra Agriculute University, Jobner Rajasthan
3. **Aneena E.R.**, Department of Community Science., College of Horticulture, Kerala Agriculture University, KAU, Thrissur, Kerala
4. **Ankur M. Arya,** Deptt. of Agricultural Engineering, Sardar Vallabhbhai Patel University of Agriculture and Technology, Meerut, Uttar Pradesh
5. **Anuj Chuadhary,** Shobhit University, Gangoh, Saharanpur, Uttar Pradesh
6. **Ashish M Mohite,** Amity Institute of Food Technology, Amity University, Noida, Uttar Pradesh
7. **B.K. Sakhale,** University Department of Chemical Technology (UDCT), Dr. Babasaheb Ambedkar Marathwada University, Aurangabad, Maharashtra
8. **Bogala Madhu**, Department of Processing and Food engineering, CTAE, Udaipur, Rajasthan
9. **Durga Shankar Bunkar,** Centre of Food Science and Technology, Institute of Agricultural Sciences, Banaras Hindu University, Varanasi, Uttar Pradesh
10. **Durvesh Kumari,** CDTRI (PCDF), Meerut, Uttar Pradesh
11. **Harsh P. Sharma**, College of Food Processing Technology and Bio-energy, AAU Anand, Gujarat
12. **Insha Zahoor,** Department of Post-Harvest Engineering & Technology, Faculty of Agricultural Sciences, Aligarh Muslim University, Aligarh, Uttar Pradesh
13. **Jaivir Singh,** Department of Agricultural Engineering, Sardar Vallabhbhai Patel University of Agriculture & Technology, Meerut, Uttar Pradesh
14. **Jitendra Kumar,** Shri Durga Ji Post Graduate College, Chandeshwar, Azamgarh-276128 Uttar Pradesh
15. **Kavindra Singh,** Department of Agricultural Engineering, Sardar Vallabhbhai Patel University of Agriculture & Technology, Meerut, Uttar Pradesh
16. **Krishna Kumar Patel,** Department of Agricultural Engineering, P.G. College Ghazipur, Uttar Pradesh
17. **Lochan Singh,** Department of Agriculture and Environmental Sciences, NIFTEM Sonepat, Haryana
18. **Mohammad Ali Khan,** Department of Post-Harvest Engineering & Technology Faculty of Agricultural Sciences, Aligarh Muslim University, Aligarh, Uttar Pradesh
19. **Namrata A. Giri**, University Department of Chemical Technology (UDCT) Dr. Babasaheb Ambedkar Marathwada University, Aurangabad, Maharashtra

20. **Neelash Chauhan,** Department of Agricultural Engineering, Sardar Vallabhbhai Patel University of Agriculture & Technology, Meerut, Uttar Pradesh
21. **Neha Sharma,** Amity Institute of Food Technology, Amity University, Noida, Uttar Pradesh
22. **P.K. Singh,** Department of A. H. & Dairying, RBS College Bichpuri Agra, Uttar Pradesh
23. **Padam Singh Champawat,** Department of Processing and Food Engineering, CTAE Udaipur, Rajasthan
24. **R.K. Goyal,** Department of A. H. & Dairying, RBS College Bichpuri Agra, Uttar Pradesh
25. **Ratnesh Kumar,** Deptt. of Agricultural Engineering, Sardar Vallabhbhai Patel University of Agriculture and Technology, Meerut, Uttar Pradesh
26. **S.K. Goyal,** Deptt. of Farm Engg. (KVK), Institute of Agricultural Sciences, BHU Barkachha, Mirzapur, Uttar Pradesh
27. **Samsher,** Deptt. of Agricultural Engineering, Sardar Vallabhbhai Patel University of Agriculture and Technology, Meerut, Uttar Pradesh
28. **Simla Thomas,** Department of Community Science, College of Horticulture, Kerala Agriculture University, KAU P.O., Thrissur, Kerala
29. **Sunil,** Deptt. of Agricultural Engineering, Sardar Vallabhbhai Patel University of Agriculture and Technology, Meerut, Uttar Pradesh
30. **Suresh Chandra,** Deptt. of Agricultural Engineering, Sardar Vallabhbhai Patel University of Agriculture and Technology, Meerut, Uttar Pradesh
31. **Tarun Kumar,** Department of Agricultural Engineering, Sardar Vallabhbhai Patel University of Agriculture & Technology, Meerut, Uttar Pradesh
32. **Upendra Singh,** Deptt of Agril Engg, Sri Karan Narendra Agriculute University Jobner Rajasthan
33. **Vaishali,** Deptt. of Agricultural Engineering, Sardar Vallabhbhai Patel University of Agriculture and Technology, Meerut, Uttar Pradesh
34. **Vijay S. Sharanagat,** Department of Food Engineering, NIFTEM, Sonepat Haryana
35. **Vikrant Kumar,** Deptt. of Agricultural Engineering, Sardar Vallabhbhai Patel University of Agriculture and Technology, Meerut, Uttar Pradesh
36. **Vipul Chaudhary,** Deptt. of Agricultural Engineering, Sardar Vallabhbhai Patel University of Agriculture and Technology, Meerut, Uttar Pradesh
37. **Vishvambhar Dayal Mudgal,** Department of Processing and Food Engineering, CTAE Udaipur, Rajasthan
38. **Yashwant Kumar Patel,** Food processing and Technology, UTD, ABVV, Bilaspur Chhatisgarh
39. **Yogesh Kumar**, Department of Food Engineering, NIFTEM, Sonepat, Haryana, India
40. **Yogesh Kumar,** Department of Plant Pathology, P.G. College Ghazipur, Uttar Pradesh

1

Physical Properties of Multi-Commodity Flour Biscuits

Suresh Chandra, Samsher and Durvesh Kumari

Abstract

Multi-commodity flours were prepared by blending rice flour, mung flour and potato flour with wheat flour in ratios of 0:0:0:100, 5:5:5:85, 10:10:10:70 and 15:15:15:55, respectively and packaged in glass jar, high density polyethylene and aluminium flexible packaging for biscuits development. The standardized formulations for biscuit had ingredients as 100 g flour/multi-commodity flour, 45 g sugar, 45 g hydrogenated fat, 1.25 g sodium bicarbonate, 1.25 g baking powder and 1.0 g curry leave powder. The biscuits were baked in convective baking oven at 180 °C for 10-15 min till baked. The well baked biscuits were removed from the oven, cooled to room temperature, packaged and stored under ambient condition. The physical quality (Mass, thickness, diameter, spread ration percent spread and bulk density), of biscuits were analyzed just after preparation under ambient condition. Results were also reported that the spread ratio of multi-commodity flour biscuits increased with decrease in the incorporation of wheat flour in multi-commodity flours. It is clear that spread ratio is mostly influenced by the diameter and thickness of biscuits.

Introduction

Bakery products are an item of mass consumption in view of its low price and high nutrient value. With rapid growth and changing eating habits of people, bakery products have gained popularity among masses. The bakery products which include bread and biscuit form the major baked foods accounting for over 82% of total bakery products produced in the country. The bakery industry in India enjoys a comparative advantage in manufacturing with abundant supply of the primary ingredients required by the industry. India is the world's second largest producer of food next to China, and has the potential of being the biggest with the food and agricultural sector. The Indian food Industry is a largest manufacturing unit for bakery, chocolates and confectionery products. India is a major manufacturing house for bakery products and is the third largest biscuit manufacturing country after USA and China. Branded, organized to unbranded, unorganized market share of biscuit has been 65% for organized sector and 35% for unorganized sector. The biscuits available in market are prepared from wheat flour (whole/refined) which lacks in good quality protein because of its deficiency in lysine; and dietary fibre contents.

Rice flour, mung flour and potato flour which are highly nutritious in protein, vitamin, minerals and lysine content has been found for its incorporation into preparation of biscuit by means of physico-chemical, functional, nutritional and sensorial evaluations. The study provides the information about a commercially viable application of increasing protein and fibre content in biscuits and also these can be solve the problem of malnutrition and other essential macro and micro nutrients deficiency among the population. The utility of rice flour, mung flour and potato flour by value addition through incorporating with wheat flour to prepare the multi-commodity flour and used to develop the biscuit and their characterization (Chandra *et al.,* 2015). Little work has been reported on development and quality assessment of biscuits made from multi-commodity flour incorporating wheat flour, rice flour, mung flour and potato flour. It is clear as per cited literature that wheat and rice flours are superior source of carbohydrate and starch content, mung flour is rich in protein content while potato flour is rich source of macro and micro mineral and vitamin contents. Thus, there is a need to develop the protein and fibre rich biscuits which would serve as nutritious products. Such basic information on physical properties should be of value not only to engineers but also food technologist, processors, bakers, and other scientists who may exploit these properties and find new uses and ideas (Chandra *et al.,* 2015)

Material and Methods

The entire research work was conducted in the Department of Agricultural Engineering and Food Technology, S.V.P. University of Agric. & Tech., Meerut (U. P.) in 2012. Multi-commodity flour was prepared by blending rice flour, mung flour and potato flour with wheat flour in ratios of 0:0:0:100, 5:5:5:85, 10:10:10:70 and 15:15:15:55 respectively and packaged in PET jar for further experiment to develop biscuits. The standardized formulations for biscuit had ingredients as 100 g flour, 45 g sugar, 45 g hydrogenated fat, 1.25 g sodium bicarbonate, and 1.25 g baking powder and 1.0 g curry leave powder. Hot liquid Hydrogenated fat and sugar were taken and creamed to a uniform consistency. The flour, required amount of water, baking powder and sodium bicarbonate were added to creamed mixture and mixed for 10 min at medium speed in dough mixer to obtain a homogeneous mixture. The dough was rolled out into thin sheet of uniform thickness and was cut into desired shape using mould. The cut pieces were placed over perforated tray and transferred into convective baking oven at 180°C for 10-15 min till baked. The well baked biscuits were removed from the oven, cooled to room temperature, packaged and stored at room temperature for further studies.

Diameter: The diameter of biscuits was measured by laying five biscuits edge to edge with the help of a scale rotating those at 90° and again measuring the diameter of five biscuits (cm) and then taking average value.

Thickness: Thickness was measured by stacking five biscuits on a top of each other and taking average thickness (cm).

Weight: Weight of biscuits (g) was measured as average of values of five individual biscuits with the help of digital electronic weighing balance.

Spread Ratio: Spread ratio was calculated by dividing the average value of diameter by average value of thickness of biscuits.

Percent spread: Percent spread was calculated by dividing the spread ratio of composite biscuit with spread ratio of control biscuits and multiplying with 100.

Bulk Density: The bulk density was determined according to the method described by Okaka and Potter (1977). Fifty (50) g sample of biscuits was put into a 100 ml graduated cylinder. The cylinder was tapped 40-50 times and the bulk density was calculated as weight per unit volume of samples.

Results and Discussions

The knowledge of important physical properties such as shape, size, volume, surface area, density, length, thickness, spread ratio, percent spread and mass of biscuits are necessary for designing of baking equipment, packaging materials, handling and storage systems. These properties are also benefitted to calculate the energy and mass balance during baking. The mass, diameter and thickness of biscuits are important to design the mild and cast of the biscuit. The effect of incorporation of flours on physical properties of freshly prepared biscuits were analyzed and discussed in following sections.

Mass: The value of mass of control (wheat flour) biscuits was lowest as compared to multi-commodity flour biscuits. The mass of biscuits was affected by the mass of dough taken for making the biscuit. The biscuits had variation in the initial weight and size of biscuit due prepared by manually. A graph plotted between variation in mass of per biscuit and type of biscuits is shown in Fig. 1. The mass per biscuit ranged 7.00 to 8.54 g. The highest mass per biscuit was measured for W_{85} biscuit (8.54g) followed by W_{70} (7.77g), W_{55} (7.02g) and lowest for wheat flour biscuit (7.00g). The study revealed that the mass of multi-commodity flour biscuits decreased with increase in the incorporation of rice, green gram and potato flour with wheat flour. The mass of biscuits decreased as the concentration of rice, green gram and potato flour increased in the multi-commodity flours. This was probably due to low OAC of rice flour and wheat flour. Similar findings were reported by Yadav *et al.,* (2012). Similar trends were found by Mridula and Wanjari (2006). They were reported that weight of biscuit decreased gradually with increase in proportion of full fat soybean flour from 5 to 20 percent and decreasing the proportion of wheat flour 100 to 80 percent. From this, it is clear that the effect of incorporation of rice, green gram and potato flour with wheat flour on mass of biscuit were found to be significant at $p<0.05$ level of significance. The study was accounted that the mass of biscuit decreased with increase in the incorporation of different flours with wheat flour. The control biscuits were observed lowest mass (7.00g) and highest for W_{85} biscuit (8.54g).

Diameter: The value of diameter of biscuits ranged 4.24 to 4.38 cm. The highest diameter was observed for W_{85} biscuits (4.38 cm) followed by W_{70} (4.34 cm) and W_{55} (4.30 cm) and lowest for control biscuits (4.24 cm). A bar chart plotted between diameter per biscuit and type of biscuits is shown in Fig. 1. It is clear that the diameter of multi-commodity flour biscuits had larger as compared to control biscuits. The diameter of multi-commodity flour biscuits decreased with increase in the incorporation of rice, green gram and potato flour with wheat flour. Diameter and spread ratio of biscuits are the important parameter used for evaluation the wheat varieties for biscuits making (Nemeth *et al.,* 1994). Larger biscuit diameter and higher spread ratio are considered as the desirable quality attributes (Yamamoto *et al.,* 1996). Similar findings were observed by Yadav *et al.,* (2012). This study revealed that diameter of biscuits decreased insignificantly with increase in the incorporation of different flours with wheat flour. The control biscuits had smaller diameter (4.24 cm) and larger for W_{85} biscuits (4.38 cm).

Thickness: The highest thickness per biscuit was measured for W_{85} biscuit (0.84 cm) followed by control biscuit (0.70 cm), W_{70} (0.66 cm) and lowest for W_{55} (0.52 cm). The physical parameter like thickness of biscuits as affected by the incorporation of different flours i.e. rice, green gram and potato flour with wheat flour are presented in Fig 1. The thickness per biscuit ranged 0.52 to 0.84 cm. From the study revealed that the thickness of biscuits decreased with increase in the incorporation of rice, green gram and potato flour with wheat flour. Highest value of thickness was observed for W_{85} biscuits as compared to control biscuits. The thickness of biscuits was influenced by the initial mass of the dough ball which was taken for the preparation of biscuits.

Decrease in diameter and thickness of multi-commodity flour biscuits with other flours with wheat flour may be due to dilution of gluten. Similar results were reported by Ajila *et al.,* (2008). Highest value of thickness was measured for W_{85} biscuits and lowest for W_{55} biscuits as compared to control biscuits.

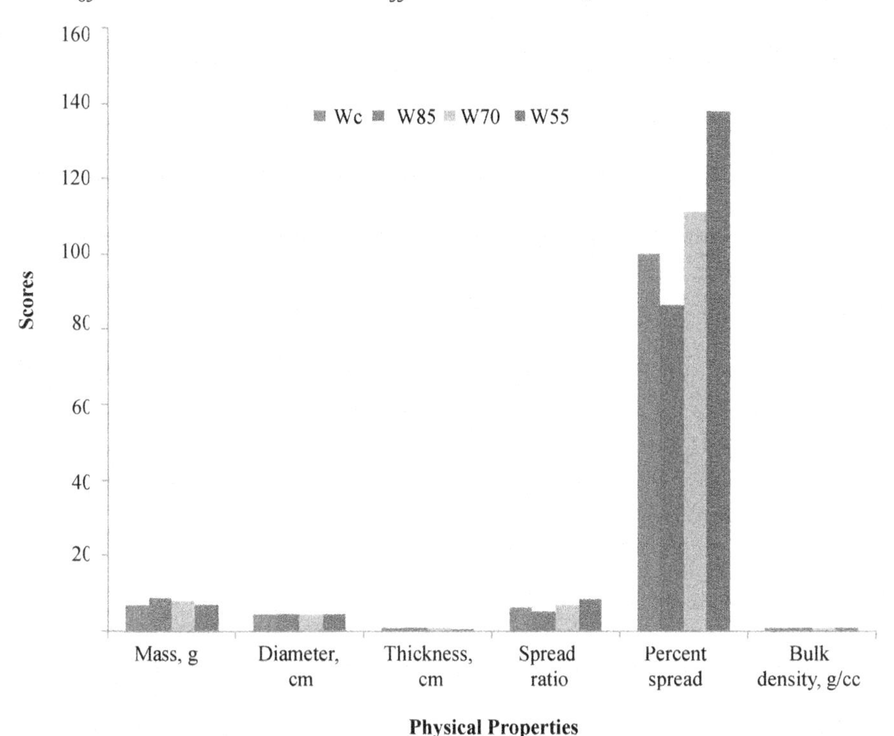

Fig. 1 Variation in Physical Properties of multi-commodity flour biscuits

Spread Ratio: The spread ratio is the ratio of diameter to thickness of biscuits. The variation in spread ratio for biscuits is given in a graph is shown in Fig. 1. The spread ratio of biscuits ranged 5.224 to 8.320. The highest spread ratio was evaluated for W_{55} biscuits (8.320) followed by W_{70} (6.676), W_{85} (5.224) and lowest for control biscuits (6.092). The results indicated that the incorporation of different flours with wheat flour increased the spread ratio of biscuits. Spread ratio of W_{85} biscuit was found lower than control biscuits but higher for W_{70} and W_{55} biscuits. Results were also revealed that the spread ratio of multi-commodity flour biscuits increased with decrease in the incorporation of wheat flour in multi-commodity flours. It is clear that spread ratio is mostly influenced by the diameter and thickness of biscuits. Spread ratio and percent spread decreased with addition of rice, green gram and potato flour. Rababah *et al.,* (2006) reported the reduction in spread ratio when chickpea, broad bean and isolate soy protein were substituted for wheat flour in biscuits. Spread ratio of other biscuit samples was found to be insignificant to each other. Spread ratio of the multi-commodity flour biscuits increased with increase in the incorporation of different flours.

Percent Spread: A graphical representation of percent spread of biscuits is shown in Fig. 1. The percent spread of biscuits varied 86.378 to 138.058. The highest score of percent spread were observed for W_{55} biscuits (138.058) followed W_{70} (111.228), control biscuits (100.00) and lowest for W_{85} biscuits (86.378). Adair *et al.,* (2001) found that mung bean paste incorporation reduced cookie spread at all the level of substitutions (25, 50, 75 and 100%) of peanut butter which was not similar to present study. Mandal *et al.,* (2004) reported that incorporation of 25 percent green gram flour in the formulation of biscuit improved height, diameter, spread ratio, colour, texture and flavour. The study revealed that the percent spread of biscuit increased with increase in the incorporation of different flours with wheat flour. The percent spread of biscuits increased with decrease in the incorporation of wheat flour. Percent spread of biscuits was influenced by the thickness and diameter of biscuits. The spread ratio of W_{55} biscuit was found to be significant as compared to control biscuits while other was found insignificant to each other. The spread ratio of biscuits was increased insignificantly with increase in the incorporation of rice, green gram and potato flour with wheat flour.

Bulk density: The variation in bulk densities of biscuits are reported in Fig. 1, which shows the effect of incorporation of rice, green gram and potato flours with wheat flour. Bulk densities of biscuits ranged 0.6782 to 0.8160 g/cc. The highest bulk density was reported for control biscuits (0.8160 (g/cc) followed by W_{85} (0.7999g/cc), W_{70} (0.7079 g/cc) and lowest for W_{55} biscuits (0.6782 g/cc). It was also noticed that level of incorporation of different flours was influenced the bulk density of biscuits. From Fig. 1, it was observed that the bulk densities among all biscuit samples decreased with increase in the incorporation of rice, green gram and potato from with wheat flour while decreased with decrease in the proportions of wheat flour in multi-commodity flours. Hence, bulk density of biscuits depends on the particle size of incorporating flours which reduced by coarse size of potato flours in biscuits. Lower density is often suggested as a quality index for biscuits (Fustier *et al.,* 2009). The bulk density of W_{70} and W_{55} biscuits were found to be significant except W_{85} as compared to control biscuits. It is clear the bulk density decreased significantly with increase in the incorporation of different flours with wheat flour. Akubor and Obiegbuna (1999) reported that bulk density of sample could be used in determining its packaging requirements as this related to the load the sample can be carry it allowed to rest directly on one another. The density is often noted as an important quality parameter in biscuit making, in particular for predicting crunchiness (Bartalucci and Launay, 2000).

Conclusions

The value of mass for control biscuits was lowest as compared to multi-commodity flour biscuits. The mass of biscuits was affected by the mass of dough taken for making the biscuit. The diameter of multi-commodity flour biscuits decreased with increase in the incorporation of rice, green gram and potato flour with wheat flour. The thickness of biscuits was influenced by the initial mass of the dough ball which was taken for the preparation of biscuits. Results were also reported that the spread ratio and percent spread of multi-commodity flour biscuits increased with decrease in the incorporation of wheat flour in multi-commodity flours. It is clear that spread ratio is mostly influenced by the diameter and thickness of biscuits. Bulk density of biscuits depends on the particle size of incorporating flours which reduced by coarse size of potato flours in biscuits.

References

Adair, M., Knight, S. and Gates, G. (2001). Acceptability of PB cookies prepared using mung bean paste as fat ingredient substitute. J. Am. Diet Assoc. 101:467–469.

Ajila, C.M., Leelavathi, K. and Rao, U.J.S.P. (2008). Improvement of dietary fibre content and antioxidant properties in soft dough biscuits with the incorporation of mango peel powder. J. Cereal Sci., 48: 319–326.

Akubor, P.I. and Obiebuna, J.E. (1999). Certain chemical and functional properties of ungerminated and germinated millet flour. J. Food Sci. Technol. 36:241–243.

Bartolucci, J.C. and Launay, B. (2000). Stress relaxation of wheat flour doughs following bubble inflation or lubricated squeezing flow and its relation to wheat flour quality. In J.D. Schofield (ed), wheat structure, biochemistry and functionality (pp. 323–331), Cambridge: The Royal Society of Chemistry.

Chandra, S., Samsher; Kumar, P., Vaishali; and Kumari, D. (2015). Effect of incorporation of rice, potato and mung flour on the physical properties of composite flours biscuits. South Asian J. Food Technol. Environ., 1(1): 64–74.

Chandra, S., Samsher; Kumari, D. (2015). Evaluation of functional properties of multi-commodity flours and sensorial attributes of composite flour biscuits. J. Food Sci. Technol., 52(6): 3681–3688.

Fustier, P., Casteigne, F. Turgeon, S.L. and Biliaderis, C.G. (2009). Impact of endogenous constituents from different flour milling streams on dough rheology and semi-sweet biscuit making potential by partial substitution of commercial soft wheat flour. LWT-Food Sci. Technol., 42:363–371.

Mandal, S., Singh, G. and Agrawal, P. (2004). Development of green gram fortified biscuits. Indian baker,3:31–38.

Mridula, D. and Wanjari, O.D. (2006). Effect of incorporation of full fat soy flour on quality of biscuits. Bev. Food World, 33(8): 35–36.

Nemeth, L.J., Williams, P.C. and Bushuk, W. (1994). A comparative study of the quality of soft wheat from Canada, Australia and United States. Cereal Foods World, 691–699.

Okaka, J.C. and Potter, N.N. (1977). Functional and storage properties of cow pea-wheat flour blends in bread making. J. Food Sci., 42: 828–833.

Rababah, T.M., Al-Mahasneh, M.A. and Ereifej, K.I. (2006). Effect of chickpea, broad bean or isolate soy protein (ISP) additions on the physic-chemical and sensory properties of biscuits. J. Food Sci., 71: 438–442.

Yadav, R.B., Yadav, B.S. and Dhull, Nisha (2012). Effect of incorporation of plantain and chickpea flours on the quality characteristics of biscuits. J. Food Sci. Technol., 49(2): 207–213.

Yamamoto, H., Worthington, S.T., Hou, G. and Ng, P.K.W. (1996). Rheological properties and baking quality of selected soft wheat in United States. Cereal Chem., 73: 215–221.

2

Assessment of Physical Properties of Food Grains

Suresh Chandra, Samsher, SK Goyal and Jitendra Kumar

Abstract

Food processing involves the transformation of raw animal or plant materials into consumer-ready products, with the objective of stabilizing food products by preventing or reducing negative changes in quality. Without these processes, we would neither be able to store food from time of plenty to time of need nor to transport food over long distances. Food processing has also gained its importance in the wide variety of diet among people throughout the globe and availability of exotic food items at various places. Processing of food items enhance the taste, flavor and aroma of the food thereby increasing the overall chances of its acceptability among the masses. So, it is essential to determine and recognize the database of physical and engineering (aerodynamic and mechanical) properties of these agricultural products because these properties play an important role in designing and developing of specific machines and their operations such as sorting, separating and cleaning, also to determine the optimum in seed metering device in pneumatic planter and precision sowing machine to suite every size of these grains.

The Engineering (physical and mechanical) properties constitute important and essential data in the design of machines, storage structures and processes. The value of this basic information is not only important to engineers but also to food scientists, processors, and other scientists who may exploit these properties and find new uses. The size and shape are, for instance, important in their electrostatic separation from undesirable materials and in the development of sizing and grading machinery (Mohsenin, 1970). The shape of the material is important for an analytical prediction of its drying behaviour. Bulk density and porosity are major considerations in designing near-ambient drying and aeration systems, as these properties affect the resistance to airflow of the stored mass. The theories used to predict the structural loads for storage structures have bulk density as a basic parameter. The angle of repose is important in designing the equipment for mass flow and structures for storage. The frictional characteristics are important for the proper design of agricultural product handling equipment (Kaleemullah and Kailappan, 2003). The major moisture-dependent physical properties of biological materials are shape and size, densities, porosity, mass of grains and friction against various surfaces. These properties have been studied for various crops such as soybean (Deshpande *et al.*,1993),pumpkin grains (Joshi *et al.*, 1993), sunflower (Gupta and Das,1997), green gram (Nimkar and Chattopadhyay, 2001), pigeon pea (Baryeh and Mangope, 2002b), black-eyed pea (Unal *et al.*, 2006), some grain legume seeds (Altuntas and Demirtola, 2007) and Faba bean (Altuntas and Yildiz,2007).

Despite an extensive search, no published literature was available on the detailed physical properties of white kidney beans and their dependency on operation parameters that would be useful for the design of processing machineries. In order to design equipment and facilities for the handling, conveying, separation, drying, aeration, storing and processing of white kidney beans, it is necessary to know their physical properties as a function of moisture content. Surface area and volume of legume seeds is an important physical characteristic in processes such as harvesting, cleaning, separation, handling, aeration, drying, storing, milling, cooking and germination (Igathinathane and Chattopadhyay, 1998). Geometric parameters of legume seeds are important for germination process as well, bigger bean seeds germinate faster than smaller and medium ones. Owing to the irregularities and variation in shapes, surface profiles and dimensions of specific food materials, it is very difficult to evaluate their actual surface areas. For food materials, such as seeds, grains, fruits or vegetables that are irregular in shape, complete specification of shape requires an infinite number of measurements.

The shapes of most natural food materials generally resemble some of the regular geometrical objects, and this feature is utilized in the theoretical estimation of the surface area utilizing certain numerical techniques. Often three measurements along the mutually perpendicular axes, namely, length, width, and thickness are used to specify the shape of the food material. The knowledge of physical properties of food materials is of importance to plant breeders, engineers, machine manufacturers, food scientists, processors, and consumers. The data on physical properties are used in designing relevant machines and equipment for harvesting, handling, transportation, separating, aeration, sizing, storing, packing and the other processing. The data have also been used for assessing the product quality.

The determination of physical properties of food materials is much complex because of their irregular shape and variability in size. Presently no single standard method is applied in determining the physical dimensions of agricultural products (Waziri and Mittal, 1983). The procedures for determination of physical properties of kidney beans are discussed below:

Grain dimensions

The average grain dimension was measured by picking 10 grains randomly. The three linear dimensions namely length (L), Width (W) and Thickness (T) were measured using a Vernier Caliper (least count 0.01mm) for all kidney beans. The measurements were taken under ambient conditions.

Diameter

Arithmetic mean diameter (AMD), Geometric mean diameter (GMD), Square mean diameter (SMD) and Equivalent diameter (EQD) of food grain can be calculated by using the following equations (Mohsenin, 1986). Geometric mean diameter (grain size) of grain is the cube root of product of three semi-axis of grain. Three major principles axes of grain were measured by Vernier callipers having least count of 0.01 mm. The grain size or GMD of grains was calculated by using by the following relationship.

$$\text{Arithmetic mean diameter(AMD)} = \frac{L + W + T}{3}$$

$$\text{Geometric mean diameter (GMD)} = (LWT)^{1/3}$$

$$\text{Square mean diameter (SMD)} = (LW + WT + TL)^{1/2}$$

$$\text{Equivalent diameter(EQD)} = \frac{\text{AMD} + \text{GMD} + \text{SMD}}{3}$$

Sphericity

Sphericity is the ratio of volume of solid to the volume of a sphere that has a diameter equal to the major diameter of the object so that it can circumscribe the solid sample. The value of sphericity ranges from 0 to 1. The sphericity (Φ) of grains was calculated by using the following relationship given by (Mohsenin, 1970):

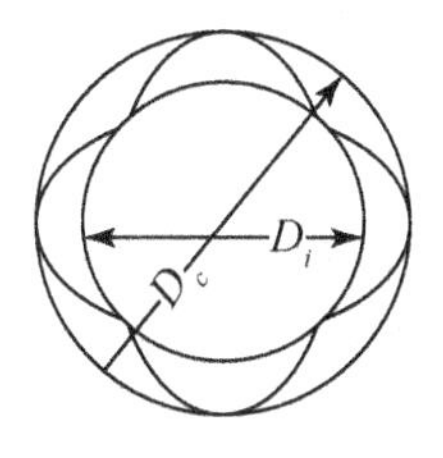

$$\text{Sphericity } (\Phi) = \frac{(LWT)^{1/3}}{L}$$

OR

$$\text{Sphericity} = \left(\frac{\text{Volume of solid sample}}{\text{Volume of circumscribed sphere}} \right)^{1/3}$$

OR

$$\text{Sphericity} = \frac{\text{geometric mean diameter}}{\text{major diameter}}$$

Another definition of sphericity is the ratio of the diameter of the largest inscribed circle (D_i) to the diameter of the smallest circumscribed circle (D_c).

$$\text{sphericity}(\Phi) = \frac{D_i}{D_c}$$

Volume and Surface Area

Major dimensions of the grain was used to calculate the volume (V) and surface area (S) of a single bean (Jain and Bal, 1997) given as below

$$\text{Volume} = \frac{\pi (GMD)^2 L^2}{6(2L - GMD)}$$

$$\text{Surface area} = \frac{\pi (GMD) L^2}{(2L - GMD)}$$

Shape Factor

Shape factor (λ) based on volume and surface area of beans was determined (McCabe and Smith, 1984) as;

$$\text{Shape factor} = \frac{a}{b}$$

Where, $a = v/w^3$, $b = s/w^2$

Thousand Kernel Mass

About one kilogram of kidney beans was divided in two equal portion 1000 kernel mass (TKM) of kidney bean was randomly picked from each portion and separately weighted using a digital electronic balance(least count 0.001 mg)

Bulk and True Density

The bulk density (ρ_b) and true density (ρ_t) are the measure of the quality of grain. Bulk density of grain is the ratio of mass of grain to its bulk volume. True density is the ratio of the mass of grain to its actual volume. The bulk volume (including pore spaces between grains) of grain is greater than actual volume (without pore spaces between grains). So, bulk density of grain is smaller than that of true density. The bulk density of beans was determined by measuring the mass of grain sample of known volume. The Beans sample was placed in a cylindrical container of volume 250 cm^3. Filling of Beans in the cylinder for density was obtained by gently tapping the cylinder vertically down to a table 30 times in the same manner in all measurements. The excess grains on the top of the cylinder were removed by sliding a string along the top edge of the cylinder. Thereafter, the mass of the grain sample was measured by an electronic balance.

True density or substance density was determined by toluene displacement method in order to avoid absorption of water during the experiment (Rahman 1995). The measurements were done at room temperature and replicated three times. The average diameter of equivalent sphere (D_e, m) was calculated in terms of mass of 1000 grain (W_{1000}, kg) and true density (ρ_t, kg/m^3) using following equation (Mohsenin, 1970).

$$De = \left(\frac{6W1000}{1000\pi\rho_t} \right)^{1/3}$$

Porosity

The total porosity (ε) was determined by using the formula (Mohsenin, 1970).

$$\varepsilon = \left(1 - \frac{\rho_b}{\rho_t} \right) \times 100$$

Specific gravity

The specific gravity bottle or pycnometer and toluene ($C_6H_5CH_3$) are used for determination of specific gravity of granular agricultural materials. The following procedure is used for determination of the specific gravity.

The weight of empty pycnometer and the weight of pycnometer filled with water at 20°C are taken. The specific gravity of toluene is determined as the ratio of the weight of toluene in the bottle and weight of distilled water in the bottle at same temperature.

$$\text{Specific gravity of toluene} = \frac{\text{weight of toluene}}{\text{weight of water}}$$

Ten gram sample of grain is placed in the pycnometer and filled with toluene to cover the sample. The air is completely exhausted from the bottle by a vacuum pump. When there is no air bubble remains in the bottle, the bottle is filled with toluene and the temperature is allowed to reach 20°C. The bottle is weighted and the specific gravity of grain is calculated by following expressions,

$$\text{Specific gravity of grain} = \frac{\text{specific gravity of toluene} \times \text{weight of grain}}{\text{weight of the toluene displaced by the grain}}$$

Shrinkage

Shrinkage is the decrease in volume of the food during processing such as drying. When moisture is removed from food during drying, there is a pressure imbalance between inside and outside of the food. This generates contracting stresses leading to material shrinkage or collapse. Shrinkage affects the diffusion coefficient of the material and therefore has an effect on the drying rate. In simple way, Shrinkage is also defined as the percent change from the initial apparent volume.

Apparent shrinkage is defined as the ratio of the apparent volume at given moisture content to the initial apparent volume of materials before processing; can be calculated as:

$$S_{\text{app}} = \frac{V_a}{V_0}$$

Where

V_a = apparent volume at a given moisture content (cm^3),
V_0 = initial apparent volume (cm^3).

References

Altuntas, E. and Demirtola, (2007). Effect of moisture content on physical properties of some grain legume seeds. N.Z.J. Crop Horticulture Science, 35(4): 523–433

Altuntas, E. and Yildiz, M. (2007). Effect of moisture content on some physical and mechanical properties of faba bean (*Vicia faba* L.) grains. J. Food Engg., 78: 174–183.

Baryeh, E.A. and Mangope B.K. (2002). Some physical properties of QP-38 variety pigeon pea. J. Food Eng., 56: 59–65.

Deshpande SD., Bal S. and Ojha TP. (1993). Physical properties of soybean, J. Agric. Engg. Res., 56: 89–98.

Gupta, R. K., and Das, S. K. (1997). Physical properties of sunflower seeds. J. Agric. Engg. Res., 66: 1–8.

Igathinathane, C. and Chattopadhyaya, P.K. (1998). On the development of ready to reckoner table for evaluation surface area of general ellipsoids based on numerical technique. J. Food Engg., 36: 233–247.

Jain, R.K. and Bal, S. (1997). Physical properties of pearl millet. J. Agric. Eng. Res., 66: 85–91.

Joshi, D.C., Das, S.K. and Mukherjee, R.K. (1993). Physical properties of pumpkin seeds. J. Agric. Eng. Res., 54(3): 219–229.

Kaleemullah, S. and Gunasekar, J.J. (2002). Moisture-dependent physical properties of arecanut kernels. Biosyst. Engg., 82 (3): 331–338.

Kaleemullah, S. and Kailappan, P. (2003). Geometric and morphometric properties of chillies. Internat. J. Food Properties, 6 (3): 481–498.

McCabe, W.L. and Smith, J.C. (1984). Unit operations of chemical engineering. 3rd edition, Mc Graw Hill book company, Japan.

Mohsenin, N.N. (1970). Physical properties of plant and animal materials. Gordon and Breach Sci. Pub, New York.

Mohsenin, N.N. (1986). Physical properties of plant animal material-II. Gordon and Breach Sci. Pub, New York.

Nimkar, P.M. and Chattopadhyay, P.K. (2001). Some physical properties of green gram. J. Agric. Eng. Res., 80 (2): 183–189.

Rahman, M.S. (1995). Food properties handbook. CRC Press, Boca Raton, FL.

Unal, H; Isik, E and Alpsoy, H.C. (2006). Some physical and mechanical properties of black eyed pea (*Vigna Unguiculata,* L.) grain. Pakistan J. Bio. Sci., 9(9): 1799–1806.

Waziri, A.N. and Mittal, J.P. (1983). Design related physical properties of selected agricultural product. Agril. Mechaniz. Asia, Africa and Latin America, 14(1): 59–61.

3

Evaluation of Physical and Frictional Properties of Green Peas

Suresh Chandra, Samsher and Anuj Chaudhary

Abstract

The knowledge of engineering (physical) properties constitutes important and essential data in the design of machines, storage structures and processes. The present investigation was carried out determine the physical parameter (diameters, mass, volume, density, sphericity, angle of repose etc.) of varieties KPS 10. The moisture content in green pea (KPS 10) was observed 411.77 % (db). The average values of dimensional and geometrical parameter like length (D_1), width (D_2), height (D_3), AMD, GMD and SMD were found 9.517 mm, 9.343 mm, 7.707 mm, 8.856 mm, 8.816 mm and 15.305 mm respectively. Mean values of mass of 1000 pea seeds, volume, bulk density and true density were found 329.501 g, 0.2 ml, 0.275 g/cm^3 and 0.304 g/cm^3 respectively. Average values of angle of repose, sphericity, porosity and moisture content were found 24.7, 0.926 %, 9.528 and 80.96%, respectively.

Introduction

Green peas (*Pisum sativum*) are considered as an important agricultural crop and an integral part of the human diet worldwide. The pea is most commonly small spherical seed or the seed-pod of the pod fruit. A pea is most commonly green, occasionally purple or golden yellow, pod-shaped vegetable, widely grown as a cold season vegetable crop. Pea is an important and highly nutritive vegetable widely grown throughout world. In India, it is grown as a winter vegetable in plains of North India and as a summer vegetable in hills. Pea is the choicest vegetable grown for shelled green seeds. It acquired a place of prominence not only in sumptuous banquets but in diets of ordinary and poor class of people also. Green peas are a good source of protein as well as other nutritional elements like vitamins, mineral and fibre. Green peas are also rich in folic acid; this acid can help prevent cardiovascular illnesses. Fiber helps in proper digestion of food; it is extremely important in fighting against cancer because fiber binds toxins to it and assists in removal of toxins from body. Green peas are also rich in iron, so these become an ideal food to feed lactating or pregnant women who often face lack of iron. Fresh green peas are very good in ascorbic acid (vitamin C). 100g of fresh pods carry 40 mg or 67% of daily requirement of vitamin C. Vitamin C is a powerful natural water-soluble anti-oxidant. Vegetables rich in this vitamin would help human body develop resistance against infectious agents and scavenge harmful, pro-inflammatory free radicals from the body. Peas contain phytosterols, especially β-sitosterol. Studies suggest that vegetableslike legumes, fruits and cereals rich in plant sterols help lower cholesterol levels inside the human body. The value of this basic information is not only important to engineers but also to food scientists, processors, and other scientists who may exploit these properties and find new uses. The size and shape are, for instance, important in their electrostatic separation from undesirable materials and in the development of sizing and grading machinery (Mohsenin, 1970). Yalcin *et al.*, (2005) evaluated the physical properties of pea seed as a function of moisture content. The average length, width and thickness were 7.80, 6.41 and 5.55 mm, respectively, at a moisture content of 10.06% dry basis (d.b.).

Material and methods

Experiments were conducted to studies the physical properties of Green Peas (*Pisum sativum*). The study was done in the Food Analysis Laboratory and Process and Food Technology, S.V.P. University of Agriculture and Technology, Meerut. Studies were conducted to evaluate the various physical properties of green pea and their dehydration at different types of drying such as sun drying, solar drying at 50°C, 60°C and 70°C hydration kinetics and functional properties of dried green pea flour.

Raw materials viz., green peas (KPS 10) was produced from local market of Saharanpur (Uttar Pradesh) for the present study. Kalash seeds private limited, Jalna (Maharastra), India produced an improved variety KPS 10 which were taken under the study purpose in the research. These seeds are having early maturity, picking can be done within 75 days of sowing, long pod size, 8-10 seeds per pod and strong against diseases. The green peas were cleaned manually to remove all foreign matter. Peas were stored in dry and cool place in ambient condition until testing. The determination of physical properties of agricultural materials is much complex because of their irregular shape and variability in size. Presently no single standard method is applied in determining the physical dimensions of agricultural products (Waziri and Mittal, 1983). The dimensions of agricultural materials are measured by ordinary measuring ruler in case of big objects like Coconut, Cabbage, Potato etc., by vernier calipers or micrometer screw gauge in case of relatively small objects like maize, peanut, common beans, groundnut etc. For very small objects like mustard, vegetable seed, they are measured by shadowgraph using overhead projector.

Dimensions: The average size of the pea, 10 peas were randomly picked and their three linear dimensions namely length (D_1), Width (D_2) and Thickness (D_3) were measured using a Vernier Caliper (least count 0.01mm). The measurements were taken at room temperature.

Mass, volume and density: Mass of fresh peas was determined using high accuracy electronic balance. The volume of peas was determined individually by water displacement method using a cylinder of 100 ml capacity. The mass and volume were expressed in g and ml respectively (1 ml = 1 cm^3). Densities for peas were calculated using the following equation:

$$\text{Density} = \frac{\text{Mass(g)}}{\text{volume}(\text{cm}^3)} \tag{1}$$

Geometrical and morphological properties: Sphericity, arithmetic mean diameters (AMD), geometric mean diameter (GMD) and square mean diameter (SMD) for peas were calculated by using the following equations as suggested by Mohsenin (1986):

$$\text{AMD} = \frac{D_1 + D_2 + D_3}{3} \tag{2}$$

$$\text{GMD} = \sqrt[3]{D_1 D_2 D_3} \tag{3}$$

$$\text{SMD} = \sqrt{D_1 D_2 + D_2 D_3 + D_3 D_1} \tag{4}$$

$$\text{Sphericity} = \left(\frac{GMD}{D_1}\right) \tag{5}$$

Where,

D_1 = major diameter (length), D_2 = intermediate diameter (width), and D_3 = minor diameter (height).

Bulk and True Density: The bulk density and true density are the measure of the quality of grain. Bulk density of grain is the ratio of mass of grain to its bulk volume. True density is the measure of the ratio of the mass of grain to its actual volume. The bulk volume (including pore spaces between grains) of grain is greater than actual volume (without pore spaces between grains). So, bulk density of grain is smaller than that of true density. The bulk density of peas was determined by measuring the mass of sample of known volume. The peas sample was placed in a cylindrical container of volume 250 cm^3. Packing of peas in the cylinder for density was obtained by gently tapping the cylinder vertically down to a table 10 times in the same manner in all measurements. The excess grains on the top of the cylinder were removed by sliding a string along the top edge of the cylinder. After the excess had been removed, the mass of the grain sample was measured by an electronic balance.

Porosity: The total porosity (ε) was determined by using the formula (Mohsenin, 1970)

$$\varepsilon = \left(1 - \frac{\rho_b}{\rho_t}\right) \times 100 \tag{6}$$

Where

ε = porosity (%)
ρ_b = bulk density (kg/m^3)
ρ_t = true density (kg/m^3).

Initial moisture content (IMC): The method recommended by Ranganna (1986) was used for determination of moisture content.

Angle of Repose: The Angle of Repose was determined by using a topless and bottomless cylinder of 10 cm diameter and 15 cm height. The cylinder was placed on a table and filled it with grams and rose slowly until it forms a cone. The diameter (D) and height (H) of cone was recorded. The angle of repose was calculated by using the formula as Kaleemullah (1992).

$$\text{Angle of Repose} = \tan^{-1}\left(\frac{2H}{D}\right) \tag{7}$$

Results and Discussion

The present investigation was carried out to determine the physical parameter (diameters, mass, volume, density, sphericity, angle of repose etc.) of varieties (KPS 10) of green peas (Table 1). Detailed results are discussed as below:

Table 1: Physical and frictional properties of green pea's variety: KPS 10.

Parameters	Range	Mean	SD
D_1 (length) mm	9.16–9.84	9.517	0.341
D_2 (width) mm	8.95–9.76	9.343	0.406
D_3 (thickness) mm	7.62–7.75	7.707	0.075
AMD (mm)	8.577–9.117	8.856	0.270
GMD (mm)	8.549–9.063	8.816	0.258
SMD (mm)	14.832–15.746	15.305	0.458
Volume(cm^3)	0.10–0.30	0.20	0.100
Sphericity	0.921–0.933	0.926	0.006

Parameters	Range	Mean	SD
M_{1000}(g)	314.222–339.716	329.501	13.480
Bulk density(g/cc)	0.273–0.277	0.275	0.002
True density(g/cc)	0.301–0.307	0.304	0.003
Porosity (%)	9.302–9.772	9.528	0.235
Moisture content, %	78.96–82.93	80.96	1.985
Angle of repose, degree	24.60–24.80	24.70	0.100

The average values of dimensional and geometrical parameter like length (D_1), width (D_2), height (D_3), AMD, GMD and SMD were found 9.517 mm, 9.343 mm, 7.707 mm, 8.856 mm, 8.816 mm and 15.305 mm respectively. Mean values of mass of 1000 pea seeds, volume, bulk density and true density were found 329.501 g, 0.2 ml, 0.275 g/cm^3 and 0.304 g/cm^3 respectively. Average values of angle of repose, sphericity, porosity and moisture content were found 24.7, 0.926 %, 9.528 and 80.96 % respectively.

References

Kaleemullah, S. (1992). The effect on moisture content on the physical properties of groundnut kernels. Topical Sciences, 32: 129–136

Mohsenin, N. N. (1986). Physical properties of plant and animal materials (end ed.). Gordon and Breach Science Publishers, New York.

Mohsenin, N.N. (1970). Physical properties of plant and animal materials (Vol.1: Physical characteristics and mechanical properties). Gordon and Breach Science Publishers, New York.

Ranganna, S. (1986). Hand book of analysis and quality control for fruits and vegetable products. Tata McGraw- Hill Publishing Ltd., New Delhi.

Waziri, A.N. and Mittal, J.P. (1983). Design related physical properties of selected agricultural products. Agricultural Mechanization in Asia, Africa and Latin America, 14: (1) 59–62.

Yalcın, I., Zarslan, C.O. and Akbas, T. (2005). Physical properties of pea (Pisum sativum) Seed. Journal of Food Engineering, 79:731–735.

4

Normal Water Soaking Depended Physical Properties of White Speckled Kidney Beans (*Phaseolus Vulgaris*)

Suresh Chandra and Samsher

Abstract

The research work was conducted on the moisture dependent physical properties of kidney bean (White speckled kidney beans) were analyzed. Length, width, thickness varied between 1.53-1.90, 0.79-0.94, 0.58-0.76 cm for kidney beans, respectively. Arithmetic mean diameter and geometric mean diameter were 0.96-1.19 and 0.88-1.11 cm^2, respectively. Volume, Sphericity and surface area ranged between 0.44 -0.87 cm^3, 0.57-0.60 and 2.91-4.68 cm^2 for white speckled kidney beans, respectively. The kidney bean was soaked in normal water with initial moisture content 14.22 % (db) upto 270 min and attained the final moisture upto 83.23 % (ab).

Introduction

Kidney bean (*Phaseolus vulgaris*) a grain legume, is one of the neglected tropical legumes that can be used to fortify cereal-based diets especially in developing countries, because of its high protein content. It is also a rich source of vitamin. As beans are a very inexpensive form of good protein, they have become popular in many cultures throughout the world. Kidney beans, also known as haricot bean, common bean, snap bean or navy bean, are valued for their protein- rich (23 per cent) seeds. Seeds are also rich in calcium, phosphorus and iron. The fresh pods and green leaves are used as vegetable. Kidney beans are also known as the chilli beans. These are dark red in colour and visually resemble the shape of a kidney. The bean paste was a vital ingredient in ointments for rheumatism, sciatica, eczema and common skin infections.

The manufacturing processes of common seed-based food products where hydration is incorporated are described (Salunkhe *et al.*, 1985). In these cases, hydration, also called soaking, reduces the required cooking time (Molina *et al.*, 1975) due to the even distribution of water inside the beans before cooking, leading to a better texture of the final product and increased nutrition value by (i) leaching the antinutrients such as tannins, phytic acid, some oligosaccharides, and trypsin inhibitors (Lestienne *et al.*, 2005); and (ii) shortening the cooking time where most nutrient degradation occurs. The effect of soaking at room temperature on reduction of antinutrient factors in lentil. As soaking time increased, the antinutritional factors of lentil seeds were reduced by a greater amount, especially for tannins (Abousamaha *et al.*,1985). (Egounlety and Aworh, 2003) reported that soaking at room temperature for a period of 12h can reduce the amount of oligosacharrides in soybean, cowpea and ground bean that relate to the flatulence by 17–35%.

The Engineering (physical and mechanical) properties constitute important and essential data in the design of machines, storage structures and processes. The value of this basic information is not only important to engineers but also to food scientists, processors, and other scientists who may exploit these properties and find new uses. The size and shape are, for instance, important in their electrostatic separation from undesirable materials and in the development of sizing and grading machinery (Mohsenin, 1970). The shape of the material is important for an analytical prediction of its drying behaviour. Bulk density and porosity are major considerations in designing near-ambient drying and aeration systems, as these properties affect the resistance to airflow of the stored mass. The theories used to predict the structural loads for storage structures have bulk density as a basic parameter. The angle of repose is important in designing the equipment for mass flow and structures for storage. The frictional characteristics are important for the proper design of agricultural product handling equipment (Kaleemullah and Kailappan, 2003). The major moisture-dependent physical properties of biological materials are shape and size, densities, porosity, mass of grains and friction against various surfaces. These properties have been studied for various crops such as soybean (Deshpande *et al.*, 1993), sunflower (Gupta and Das, 1997), green gram (Nimkar and Chattopadhyay, 2001), pigeon pea (Baryeh and Mangope, 2002), black-eyed pea (Unal *et al.*, 2006), some grain legume seeds (Altuntas and Demirtola, 2007) and Faba bean (Altuntas and Yildiz, 2007), kidney beans (Singh and Chandra, 2014)

Despite an extensive search, no published literature was available on the detailed physical properties of white speckled kidney beans and their dependency on operation parameters that would be useful for the design of processing machineries. Surface area and volume of legume seeds is an important physical characteristic in processes such as harvesting, cleaning, separation, handling, aeration, drying, storing, milling, cooking and germination (Igathinathane and Chattopadhyay, 1998). In order to design equipment and facilities for the handling, conveying, separation, drying, aeration, storing and processing of white kidney beans, it is necessary to know their physical properties as a function of moisture content. Owing to the irregularities and variation in shapes, surface profiles and dimensions of specific food materials, it is very difficult to evaluate their actual surface areas. For food materials, such as seeds, grains, fruits or vegetables that are irregular in shape, complete specification of shape requires an infinite number of measurements.

Material and Methods

The investigation was undertaken at the Food Analysis Laboratory and Process and Food Engineering laboratory in Department of Agricultural Engineering and Food Technology, S.V.P. University of Agriculture and Technology, Meerut (India) in 2012-13. Studies were conducted to evaluate the physical properties white speckled kidney bean (WSKB) was procured from local market for the present study. The beans were cleaned manually to remove all foreign matter such as chaff, dust and stones. They were stored in dry and cool place in ambient condition until further study. The determination of physical properties of food materials is much complex because of their irregular shape and variability in size. Presently no single standard method is applied in determining the physical dimensions of agricultural products (Waziri and Mittal, 1983). The procedures for determination of physical properties of kidney beans are discussed below:

Moisture content: Initial moisture content of samples was determined by hot air oven drying method as recommended by AOAC (2000).

Grain dimensions: The average grain dimension was measured by picking 10 grains randomly. The three linear dimensions namely length (L), Width (W) and Thickness (T) were measured using a Vernier Caliper (least count 0.01mm) for all kidney beans. The measurements were taken under ambient conditions. Arithmetic mean diameter (AMD), Geometric mean diameter (GMD), Square mean diameter (SMD) and Equivalent diameter (EQD) of kidney beans were calculated by using the following equations (Mohsenin, 1986). Geometric mean diameter (grain size) of grain is the cube root of product of three semi-axis of grain. Three major principles axes of grain were measured by Vernier callipers having least count of 0.01 mm. The grain size or GMD of grains was calculated by using by the following relationship.

$$\text{AMD} = \frac{L + W + T}{3} \quad (1)$$

$$\text{GMD} = (\text{LWT})^{1/3} \quad (2)$$

$$SMD = (LW + WT + TL)^{1/2} \quad (3)$$

$$EQD = \frac{AMD + GMD + SMD}{3} \tag{4}$$

Sphericity: The sphericity (Φ) of grains was calculated by using the following relationship given by (Mohsenin, 1970):

$$\text{Sphericity } (\Phi) = \frac{(LWT)^{1/3}}{L} \tag{5}$$

Volume and Surface Area: Major dimensions of the grain was used to calculate the volume (V) and surface area (S) of a single bean (Jain and Bal, 1997) given as below

$$\text{Volume} = \frac{\pi(GMD)^2L^2}{6(2L - GMD)} \tag{6}$$

$$\text{Surface area} = \frac{\pi(GMD)L^2}{(2L - GMD)} \tag{7}$$

Results and Discussion

The study was undertaken to sdtudy the effect of normal water soaking on Physical Properties of white speckled Kidney Beans (*Phaseolus vulgaris*). Physical properties were evaluated on soaking in normal water and regular interval of 30 minutes. The following physical properties i.e. axial dimension viz. Length (L), Width (W), Thickness (T), AMD, GMD, SMD, EQD and Moisture Content, Volume, Surface area and sphericity.

Grain Dimension

All three axial dimensions increased with increased in moisture content from 14.22 to 83.23% (d.b.). The range of mean dimension of a grain measured at above moisture content range was: length 1.53 - 1.90 cm, width 0.79-0.94 cm and thickness 0.58-0.76 cm represent in Fig. 1. The average diameter calculated by the Arithmetic mean and Geometric mean are also presented in Table 1 and plotted in Fig. 1. The AMD, GMD, SMD and EQD ranged from 0.96 to 1.19 cm, 0.88 to 1.11, 1.16 to 1.98 cm and 1.15 to 1.43 cm, respectively with increasing moisture content range from 14.22 to 83.23% (d.b.). The results showing increase in grain size with increase in moisture content were in agreement with the earlier findings for green gram (Nimkar and Chattopadhyay 2001), lentil seed (Amin *et al*., 2004), fenugreek seeds (Altuntas *et al.,* 2005), kidney bean (Isik and Unal, 2007), feba bean (Altuntas and Yildiz, 2007), lathyrus (Zewdu and Solomon, 2008) and for soybean (Kibar and Ozturk, 2008). The following regression equations were developed for length, width, thickness, AMD and GMD with moisture content (m.c. % dry basis)

Length (cm),	$L = 0.0042M_c + 1.542$	($R^2 = 0.853$)
Width (cm),	$W = 0.001M_c + 0.825$	($R^2 = 0.326$)
Thickness (cm),	$T = 0.001M_c + 0.606$	($R^2 = 0.662$)
Arithmetic Mean Diameter (cm),	$\text{AMD} = 0.002M_c + 0.988$	($R^2 = 0.738$)
Geometric Mean Diameter (cm),	$\text{GMD} = 0.002M_c + 0.909$	($R^2 = 0.701$)
Square mean diameter (cm),	$\text{SMD} = 0.003Mc + 1.65$	($R^2 = 0.696$)
Equivalent Diameter (cm),	$\text{EQD} = 0.002Mc + 1.175$	($R^2 = 0.749$)

Volume of the grain

The volume of a grain increased from 0.44 to 0.87 cm^3 with increasing moisture content and soaking period represented in Table 1 and Fig.2. The absorption of water by the grain affected the length, width and thickness of grain. The increment in dimensions, it also increased the volume of the grains. This results are in conformity with the results of reported for gram (Dutta *et al.,* 1988), soybean (Despande *et al.,* 1993; Kibar and Ozturk, 2008), lathyrus (Kenghe *et al.,* 2013). The following general expression can be used to describe the relation between moisture content and grain volume:

$$\text{Volume (V)} = 0.004\ M_c + 0.470\ (R^2 = 0.763)$$

Surface area of grain

The surface area of the white speckled kidney bean (WSKB) grain increased with increase in soaking time and moisture content. The value was varies 2.91 to 4.68 cm^2 with increasing soaking period up to 240 min (Table 1). The variation of the surface area with the moisture content of WSKB grain is shown in Fig. 2. Similar trend has been reported for linseed (Selvi et al., 2006), red kidney bean grains (Isik and Unal, 2007) and for lathyrus (Kenghe *et al.,* 2013). The relationship between moisture content and surface area can be mathematically represented as:

$$\text{SA} = 0.0.018\ M_c + 3.10\ (R^2 = 0.765)$$

Table 1: Physical Properties of WSKB at normal water soaking

Time (min.)	MC (db)	L (cm)	W (cm)	T (cm)	AMD (cm)	GMD (cm)	SMD (cm)	EQD (cm)	Vol. (cm3)	SA (cm2)	SPR (Φ)
0	14.22	1.53	0.79	0.58	0.96	0.88	1.16	1.15	0.44	2.91	0.60
30	23.51	1.65	0.83	0.63	1.04	0.95	1.71	1.23	0.55	3.46	0.57
60	29.50	1.66	0.86	0.68	1.06	0.99	1.77	1.27	0.62	3.77	0.59
90	37.36	1.77	0.93	0.71	1.14	1.05	1.88	1.34	0.73	4.17	0.59
120	46.36	1.77	0.93	0.74	1.15	1.07	1.91	1.37	0.75	4.24	0.60
150	57.21	1.79	0.93	0.72	1.15	1.06	1.90	1.37	0.75	4.23	0.59
180	68.24	1.79	0.91	0.72	1.14	1.05	1.89	1.36	0.73	4.20	0.59

(Contd.)

Time (min.)	MC (db)	L (cm)	W (cm)	T (cm)	AMD (cm)	GMD (cm)	SMD (cm)	EQD (cm)	Vol. (cm3)	SA (cm2)	SPR (Φ)
210	75.05	1.80	0.93	0.72	1.15	1.06	1.91	1.37	0.76	4.27	0.59
240	80.62	1.87	0.84	0.72	1.16	1.06	1.87	1.36	0.77	4.40	0.57
270	83.23	1.90	0.94	0.76	1.19	1.11	1.98	1.43	0.87	4.68	0.58

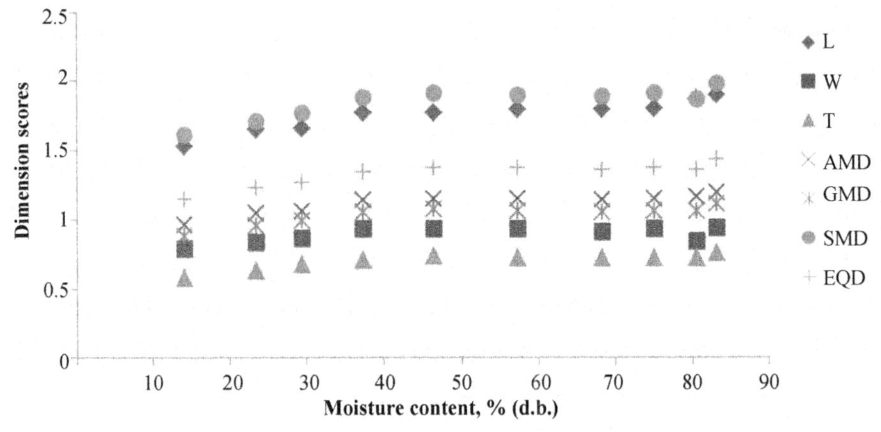

Fig. 1: Grain dimensions of WSKB at normal water soaking3

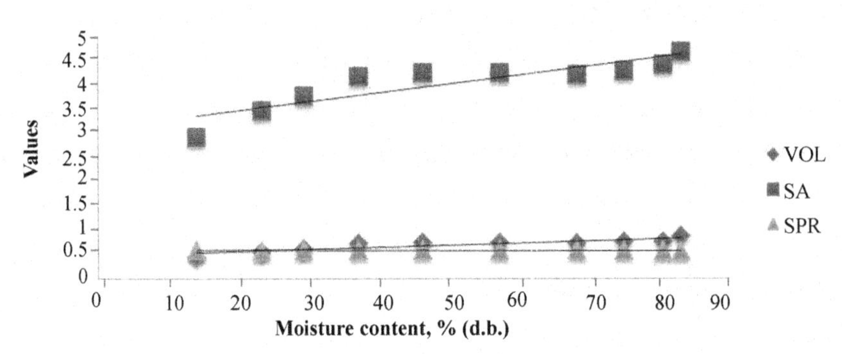

Fig.2 Physical properties of WSKB at normal water soaking

Sphericity

The value of sphericity was calculated individually by using the data on GMD and the major axis of the grain and the result obtained are presented in Table 1 and Fig.2. The result indicated that the sphericity of the grain increased from 0.57 to 0.60 in the specified moisture level as shown in Table 1. The initial increase of sphericity could due to relatively proportional increase in length, width and thickness. However, beyond second moisture content there was relatively greater increase in length as compared to width and thickness which might probably resulted in slight reduction in sphericity (Zewdu and

Solomon, 2008). An initial increase in sphericity upto 20% of moisture content and a decrease with further increase in moisture content was reported for okra seed (Sahoo and Srivastava, 2002). This indicates that different grains might behave differently in terms of the relative changes in length, width and thickness which could affect sphericity. The variation of moisture content and volume can be expressed mathematically as follows:

$$\text{Sphericity} = -0.000M_c + 0.594 \; (R^2 = 0.102)$$

References

Abousamaha, O.R., Elmahdy, A.R., Moharram, Y.G. (1985). Effect of soaking on the quality of lentil seeds. Zeitschrift Fur Lebensmittel-Untersuchung Und-Forschung, 180(6): 485–490.

Altuntas, E. and Demirtola (2007). Effect of moisture content on physical properties of some grain legume seeds. N.Z.J. Crop Horticulture Science, 35(4): 523–433

Altuntas, E. and Yildiz, M. (2007) Effect of moisture content on some physical and mechanical properties of faba bean (*Vicia faba* L.) grains. Journal of Food Engineering, 78: 174–183.

Altuntas, E., Ozgoz, E. and Taser, O.F. (2005). Some physical properties of fenugreek (Trigonella foenum-graceum L.) seeds. Journal of Food Engineering, 71: 37–43.

Amin, M.N., Hossain, M.A. and Roy, K.C. (2004). Effect of moisture content on some physical properties of lentil seeds. Journal of Food Engineering, 65: 83–87.

AOAC (2000). Official methods of analysis of AOAC International (17th ed.). Gaitherburg. USA: AOAC International Inc.

Baryeh, E.A. and Mangope, B.K. (2002). Some physical properties of QP-38 variety pigeon pea. Journal of Food Engineering, 56: 59–65.

Deshpande, S.D., Bal, S. and Ojha TP (1993). Physical properties of soybean. Journal of Agricultural Engineering Research, 56(2): 89–98

Dutta, S.K., Nema, V.K. and Bhardwaj, R.K. (1988) Physical properties of gram. Journal of Agricultural Engineering Research, 39: 259–268.

Egounlety, M. and Aworh, O.C. (2003). Effect of soaking, dehulling, cooking and fermentation with Rhizopus oligosporus on the oligosaccharides, trypsin inhibitor, phytic acid and tannins of soybean (*Glycine max* Merr.), cowpea (*Vigna unguiculata* L. Walp) and groundbean (*Macrotyloma geocarpa* Harms). Journal of Food Engineering, 56(2–3): 249–254.

Gupta, R.K. and Das, S.K. (1997). Physical properties of sunflower seeds. Journal of Agricultural Engineering Research, 66: 1–8.

Igathinathane, C. and Chattopadhyaya, P.K. (1998). On the development of ready to reckoner table for evaluation surface area of general ellipsoids based on numerical technique. Journal of Food Engineering, 36: 233–247.

Isik, E. and Unal, H. (2007) Moisture dependent physical properties of white speckled red kidney bean grains. Journal of Food Engineering, 82(2):209–216.

Jain, R.K. and Bal, S. (1997). Physical properties of pearl millet. J. Agric. Eng. Res., 66: 85–91.

Kaleemullah, S. and kailappan, P. (2003). Geometric and morphometric properties of chillies. International Journal of Food Properties. 6(3): 481–498.

Kenghe, R.N., Nimkar, P.M. and Shirkole, S.S. (2013). Moisture dependent physical properties of lathyrus. Journal of Food Science and Technology, 50(5): 856–867.

Kibar H, Ozturk T (2008) Physical and mechanical properties of soybean. International Agrophysics, 22:239–244.

Lestienne, I., Icard-Verniere, C., Mouquet, C., Picq, C. and Treche, S. (2005). Effects of soaking whole cereal and legume seeds on iron, zinc and phytate contents. Food Chemistry 89(3): 421–425.

Mohsenin, N.N. (1970). Physical properties of plant and animal materials. Gordon and Breach Science Publishers, New York

Mohsenin, N.N. (1986). Physical properties of plant animal material-II. Gordon and Breach Sci. Pub, New York.

Molina, M.R., Delafuente, G. and Bressani, R. (1975). Interrelationships between storage, soaking time, cooking time, nutritive value and other characteristics of the black bean (*Phaseolus vulgaris*). Journal of Food Science, 40(3): 587–591.

Nimkar, P.M. and Chattopadhyay, P.K., (2001). Some physical properties of green gram. Journal of Agricultural Engineering Research, 80 (2): 183–189.

Sahoo, P.K. and Srivastava, A.P. (2002) Physical properties of okra seed. *Biosystem Engineering*, 83(4):441–448.

Salunkhe, D.K., Kadam, S.S. and Chavan, J.K. (1985). Post harvest biotechnology of food legumes. CRC Press, Boca Raton, pp 160.

Selvi CS, Pinar Y, Yesiloglu E (2006). Some physical properties of linseed. Biosystem Engineering, 8: 7–8.

Singh, Y. and Chandra, S. (2014). Evaluation of physical properties of kidney bean (*Phaseolus vulgaris*). Food Science Research Journal, 5(2):125–129

Unal, H; Isik, E and Alpsoy, H.C. (2006). Some physical and mechanical properties of black eyed pea (*Vigna Unguiculata*, L.) grain. Pakistan Journal of Biological Science, 9(9):1799–1806.

Waziri, A.N. and Mittal, J.P. (1983). Design related physical properties of selected agricultural product. Agricultural Mechanization-Asia, Africa and Latin America, 14(1):59–61.

Zewdu, A. and Solomon, W. (2008) Moisture dependent physical properties of grass pea (*Lathyrus sativus* L.) seeds. International Agricultural Engineering (CIGR e-journal) Manuscript, FP 06 027 10, 102–109.

5

Cereal Grains: Frictional and Aerodynamic Properties

Vikrant Kumar, Jaivir Singh, Suresh Chandra, Neelash Chauhan Ratnesh Kumar, Sunil and Vipul Chaudhary

Abstract

The frictional properties such as coefficient of friction and angle of repose are very important in designing of storage bins, hoppers, chutes, conveyors, harvesters and threshers etc. Angle of repose is the angle made with the horizontal at which the material will stand when piled. The angle of repose is also important in designing the equipment for mass flow and structures for storage. The static angle of repose (β) was determined using an open-ended cylinder of 150 mm diameter and 250 mm height. The higher the moisture content, the higher the cohesion between the seeds. In terms of flow ability, the seeds are heavier and the inertia to move is increased. Coefficient of internal friction is the frictional resistance of seeds with each other. Coefficient of external friction is the frictional resistance between the seeds and the surface. The rolling resistance is directly proportional to the weight of the rolling object and to the coefficient of rolling resistance which is dependent on the rigidity of the supporting surface and indirectly proportional to the effective radius of the rolling object. The air velocity at which an object remains in a suspended state in a vertical pipe under the action of the air current is called terminal velocity of the object.

Introduction

Friction is a set of phenomena observed at the point of contact between two materials. It leads to the loss of energy between moving objects and the wear of surfaces that come into contact, which increases surface temperature and can even produce acoustic effects. The processes that occur between a particle and the friction surface are termed as external friction, whereas internal friction takes place between particles (Grochowicz, 1994). The coefficient of friction is associated with angle of internal friction or angle of repose which helps us in predicting stability of slope in design of conveyor or design of hopper for regulating feed rate. The coefficient of friction for different slopes is determined for all three surfaces (Singh, *et al.*, 2016).

A thorough understanding of frictional forces is required for analyzing and modeling various processes. The knowledge of frictional properties is essential for the selection of sowing, harvesting, transport, cleaning, sorting, storage and processing parameters of plant materials (Horabik, 2001; Altuntas, 2007; Kabas *et al.*, 2007; Kram, 2008; Riyahi *et al.*, 2011; Jouki, 2012 and Sologubik *et al.*, 2013). Biological materials are characterized by morphological variation, and their frictional properties can differ significantly.

Coefficient of friction is the tangent of dynamic angle of repose. It is estimated with different types of surfaces including, glass, plywood, galvanized iron, wood and stainless steel. The maximum values of coefficient of friction were obtained on the surface of wood and the minimum were obtained on the glass also, the surface of the stainless steel gives the second lowest values of coefficient of friction and it is recommended to use this material in the structure of seed hopper in planters, silos and storage containers (El Fawal, 2009).

Frictional properties

The frictional properties such as coefficient of friction and angle of repose are very important in designing of storage bins, hoppers, chutes, conveyors, harvesters and threshers etc.

In mechanical and pneumatic conveying system, the materials generally moves or slides in direct contact with the through, casing and other components of the machine. Thus various parameters affect the power requirement to drive the machine. Among these parameters the frictional losses one of the factors which must be overcome by providing additional power to the machine. Hence, the knowledge of frictional properties of the agricultural materials is necessary; therefore, some of the important frictional properties of agricultural products have been described below.

Angle of Repose: Angle of repose is the angle made with the horizontal at which the material will stand when piled. The angle of repose is also important in designing the equipment for mass flow and structures for storage (Davies, 2009). The angle of repose was determined with a help of a cylinder arrangement. The cylinder was filled with the seeds up to the top and then slowly lifted, thus a conical heap was formed. The Size, Shape, moisture content and orientation of the cereal seeds affect the angle of repose (Pawanpreet and Singh 2016). The angle of some food grains is given in Table 1. The height and the base diameter of the heap were measured. The angle of repose was calculated as:

$$\alpha = \tan^{-1}\left[\frac{2H}{D}\right]$$

Where,

α = Angle of Repose.
H = Height of the cone (cm), and
D = Base diameter of the cone (cm).

Types of angle of repose: There are two types of angle of repose, (1) Static Angle of Repose and (2) Dynamic Angle of Repose.

Static Angle of Repose: The static angle of repose is the angle of friction taken up by cereal seeds to just slide upon it. The static angle of repose (β) was determined using an open-ended cylinder of 150 mm diameter and 250 mm height. The cylinder was placed at the center of a circular plate with diameter of 350 mm. It was filled with the samples until a cone formed on the circular plate. The diameter (D) and height (H) of the cone was recorded. The filling angle of repose was calculated using the following formula (Mohsenin, 1986; Khodabakhshian *et al.*, 2017).

$$\beta = \arctan\left[\frac{2H}{D}\right]$$

Dynamic angle of repose: A plywood box of 300 × 300 × 300 mm^3 with a removable front panel was used to measure the dynamic angle of repose (θ). The box was filled with the samples, and then the front panel was quickly removed, allowing the samples to flow out and assume a natural slope. The empting angle of repose was computed using the vertical depth and radius of spread values (Ozarslan, 2002; Razavi *et al.*, 2007; Khodabakhshian *et al.*, 2017).

Table 1: Angle of repose of some grains

Grains	Angle of repose, degree
Wheat	23–28
Paddy	30–45
Maize	30–40
Barley	28–40
Millets	20–25
Rye	23–28

Source: (Sahay and Singh 1994).

The filling angle of repose increased linearly as maturity increased. The empting angle of repose increased linearly with an increase in maturity stage (Khodabakhshian *et al.*, 2017). The emptying angle of repose assumed higher values than the filling angle of repose for cereal grains at all maturity stages.

Effect of moisture content on angle of repose

The higher the moisture content, the higher the cohesion between the seeds. In terms of flow ability, the seeds are heavier and the inertia to move is increased. This increase in resistance to flow prevents seeds from sliding on each other, thereby increasing the angle of repose of the seeds. (Nimkar and Chattopadhyay 2001; Tavakoli, *et al*., 2009; & Zareiforoush, *et al*., 2009) all suggested a linear increase too for green gram seeds, *Parkiafillicoidea* specie of locust bean, barley grains and paddy grains respectively. It has been found that the angle of repose increase with the increase of moisture content of materials. This variation of angle of repose with moisture content occurs because surface layer of moisture surrounding the particle holds the aggregates of grain together by the surface tension (Sahay and Singh 1994).

The angle of repose of mungbean seed increased from 31.66% to 40.33% with the increase in moisture content from 8.72% to 27.41% (d.b). The variation in thousand seed mass for seeds with moisture content was significant ($p < 0.01$). It seems that at higher moisture content seed might tend to stick together due to the plasticity effect (stickiness) over the surface of the seed resulting in better stability and less slidity thereby increasing the angle of repose (Irtwange and Igbeka 2002). Singh and Goswami (1996), Nimkar and Chattopadhyay (2001), Amin *et al*., (2004) reported a increase in angle of repose with increase in the moisture content for cumin seed, green gram, millet, lentil and fenugreek, respectively. The angle of repose (α) of seed was found to bear the following relationship with moisture content (Mehran *et al*., 2010).

$$\alpha = - 0.0068\ Mc^2 + 0.7018\ Mc + 26.133$$

Angle of Internal Friction

The angle of internal friction is a measure of the ability of a unit of rock or soil to withstand a shear stress. It is the angle, measured between the normal force and resultant force that is attained when failure just occurs in response to a shearing stress. The angle of internal frictional is an important property which helps to estimate the lateral pressure in storage silos. Angle of internal friction values are also used in designing of storage bins and hopper for gravity discharge. The coefficient of friction between grains is required as a design parameter for design of shallow and deep bins (Sahay and Singh 1994).

Coefficient of Internal Friction

Coefficient of internal friction is the frictional resistance of seeds with each other. A small wooden box known as cell of dimension 10 cm × 10 cm × 3 cm was put into a larger wooden box known as a guided frame of dimensions 20 cm × 16 cm × 3.5 cm. The cell was tied with a fine copper wire attached to a fixed pulley. The other end of a copper wire was tied to a pan. The weight (W_1) was put on the pan so as to just slide the cell and the box together. Thereafter a fixed amount of sample of seeds of about 100 g (W) was put into the empty box and the cell was again placed into the box. The weight required to just slide the filled box and cell (W_2) was put on the pan (Pawanpreet and Singh 2016). The coefficient of internal friction was calculated as:

$$\mu_i = \frac{W_2 - W_1}{W}$$

Where,

μ_i = Coefficient of internal friction.
W_1 = Weight required to slide empty box and cell.
W_2 = Weight required to slide filled box and cell.
W = Weight of seeds.

Coefficient of external friction

Coefficient of external friction is the frictional resistance between the seeds and the surface. A box of 20 cm × 16 cm × 3.5 cm was placed on a plywood surface and was tied to a copper wire. The copper wire was attached to the box. The wire was mounted on a pulley and the other end of the wire was attached to a pan. The weight (W_1) was put on the pan so as to just slide the cell and the box together. Thereafter a fixed amount of sample of seeds of about 100 g (W) was put into the empty box and the cell was again placed into the box. The weight required to just slide the filled box and cell (W_2) was put on the pan. The same procedure was followed for sliding on a GI sheet (Pawanpreet and Singh 2016). The coefficient of external friction was calculated as:

$$\mu_e = \frac{W_2 - W_1}{W}$$

Where,

μ_e = Coefficient of external friction.
W_1 = Weight required to slide empty box.
W_2 = Weight required to slide filled box.
W = Weight of seeds.

Difference between angle of repose and angle of internal friction

The relation between angle of repose and angle of internal friction in a fragmented mass is discussed, and experiments with crushed rocks are reported. It is concluded that the angle of repose is not generally the angle of internal friction of the material in the pile, but of the same material in a more closely packed condition. It approximates to the angle of solid friction of the material upon itself (Metcalf, 1966).

Note: The angle of repose is generally higher than angle of the internal friction for the grains of approximately the same moisture content and density.

Static Friction

The friction may be defined as the frictional forces acting between surfaces of touching at slow down with regard to each other. The angle of static friction is measured by placing the analyzed material on a horizontally adjusted friction plate and by slowly increasing the angle of inclination until the seed begins to move (Grochowicz, 1994; Jouki and Khazaei, 2012; Kabas *et al*., 2007; Riyahi *et al*., 2011 and Zdzislaw, 2013).

Coefficient of static friction

Friction is a force applied on the object to oppose the motion. The formula of coefficient of static friction is obtained from the static friction force. The coefficient of static friction μ_s is determined with the use of the below formula:

$$\mu_s = \tan \alpha$$

or

$$\mu_s = \frac{F_s}{N}$$

Where,

μ_s = coefficient of static friction.
F_s = Static Friction Force.
N = Normal Force.

Effect of moisture content on static coefficient of friction

The plots of static coefficient of friction of mungbean seeds obtained experimentally on four structural surfaces against moisture content in the range of 8.72–27.41% (d.b.) are shown in Fig. 1. It was observed that the static coefficient of friction increased with increase in moisture content for all the surfaces ($p < 0.05$). It seems that at higher moisture contents the seed became rougher and its sliding characteristics were diminished, therefore the coefficient of static friction increased. Also due to increasing the stickiness and adhesion between seeds, and material surfaces at higher moisture contents, the resulting adhesive force plays an important role in increasing the value for the coefficient of static friction.

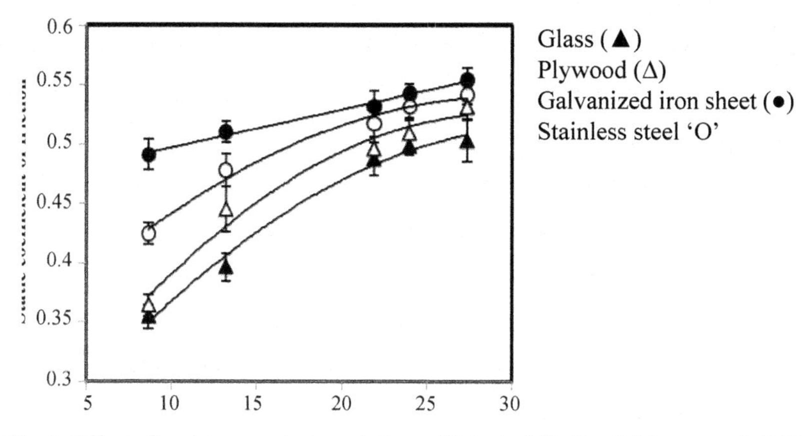

Fig. 1 Effect of moisture content on static coefficient of friction (Mehran, *et al*., 2010).

Kinetic Friction

The kinetic friction may be defined as the friction acts between the moving surfaces. The magnitude of the force depends on the coefficient of kinetic friction between the two kinds of material.

Coefficient of kinetic friction

The coefficient of friction is defined as the ratio of force of friction force to the normal force. The coefficient of kinetic friction μ_k is determined with the use of the below formula:

$$\mu_k = \frac{F}{N}$$

Where,

μ_k = coefficient of kinetic friction.

F = Kinetic Force.

N = Normal force.

Rolling Resistance or Sliding motion

The friction surface in sliding motion and rolling motion is a combination of both. The coefficient of kinetic friction was determined based on time t required for traveling the distance of S = 140 mm on a plane inclined at angle α_s (Grochowicz, 1994). The rolling resistance is directly proportional to the weight of the rolling object and to the coefficient of rolling resistance which is dependent on the rigidity of the supporting surface and indirectly proportional to the effective radius of the rolling object.

$$\mu_k = \tan\alpha_s - \frac{2S}{gt^2 \cos\alpha s}$$

Coefficient of rolling resistance

The coefficient of rolling resistance and friction is indication of how great the rolling resistance is for a given normal force between the wheel and the surface upon which it is rolling. For rolling seeds, the coefficient of rolling friction f was calculated from the below equation (*Lawrowski*, 2008; *Nosal*, 2012).

$$F = \mu_t \cdot r$$

Where, r is the radius of a rolling seed.

Based on the expected seed distribution on a given friction surface, it was assumed that radius represents the average half thickness (T) and length (L) of a given seed and it equals:

$$r = \frac{T + L}{4}$$

Various mechanisms have been designed by making use of difference in rolling resistance of the materials. One such example is the separation of potatoes and stones. The rolling resistance of stones differs from that of potatoes. The equal rolling resistance is observed only in the case of large potatoes and very small stones (*Sahay* and *Singh*, 1994).

Aerodynamic properties

The Aerodynamic properties of agricultural products are necessary for designing air and separating systems. Terminal velocity was measured by using a seed blower (air column system). For each experiment, a seed was dropped into the air stream from the top of the air column, in which air was blown to suspend the seed in the air stream. The air velocity near the location of the seed suspension was measured by a hot wire anemometer having a least count of 0.01 m/s (Pandiselvam *et al.*, 2013).

Terminal velocity (V_t): The air velocity at which an object remains in a suspended state in a vertical pipe under the action of the air current is called terminal velocity of the object (Pandey, 1998). The terminal velocities are useful for air conveying or pneumatic separation of materials in such a way that when the air velocity is greater than the terminal velocity, it lifts the particles. The air velocity at which the seed remains in suspension is considered as terminal velocity (Kachru, *et al.*, 1994). Moreover, the information of fracture characteristics of seed is imperative for a rational design of efficient grinding systems, as well as the optimization of the process and product parameters. In addition, physical and mechanical characteristics vary extensively with seed moisture content and are important to the process/product designer. (Bahnasawy, 2007; Gharibzahedi, *et al.*, 2009). The equation below was then used to determine the terminal velocity of the seed and kernel as given by (Mohsenin, 1986; Nalbandi *et al.*, 2010; Igwillo, 2017). The terminal velocity can be calculated by following formula:

$$V_t = \left[\frac{2W\ (\rho_p - \rho_f)}{C\ A_p \rho_p \rho_f}\right]^{1/2}$$

Where,

V_t = Terminal Velocity (m/s).

W = Weight of particle (kg).

ρ_t = density of particle (kg/m).

ρ_f = density of air (1.1644kg/m at room temperature).

C = Drag coefficient (dimensionless).

A_p = projected area of particle (m^2).

Effect of moisture content on terminal velocity

The values of the terminal velocity for seeds at different moisture levels varied from 4.96 to 5.81 m/s ($P < 0.01$) are show in Fig. 2. The increase in terminal velocity with increase in moisture content within the range studied can be attributed to the increase in mass of an individual seed per unit frontal area presented to the air stream. Singh and Goswami (1996), Suthar and Das (1996), Nimkar and Chattopadhyay (2001), Gezer, *et al.* (2002), Konak, *et al.* (2002), Sacilik, *et al.* (2003) and Mehran, *et al.*, (2010) have reported a increase in terminal velocity with increase in the moisture content for cumin seed, karingda seed, green gram, apricot kernel, chick pea seed and hemp seed, respectively. The relationship between moisture content and terminal velocity (V_t) can be represented by the following equation:

$$V_t = -\ 0.0011\ Mc^2 + 0.0907\ Mc + 4.2182$$

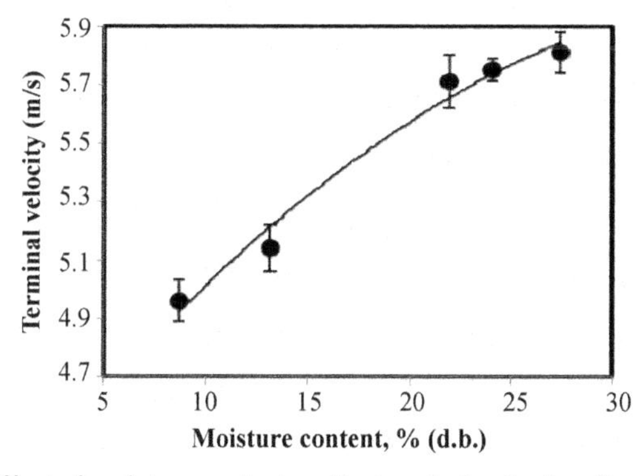

Fig. 2 Effect of moisture content on the terminal velocity of mungbean seed (Mehran *et al.*, 2010).

Reynolds number (R)

Reynolds number was calculated using the equation given by (Nwigbo, *et al.*, 2008; Igwillo, 2017).

$$R_e = \frac{DV_t \rho_f}{\mu_f}$$

Where,

R_e = Reynolds number (dimensionless).
D = arithmetic mean diameter of particle (m).
V_t = terminal velocity (m/s).
ρf = mass density of air (1.1644 kg/m at room temperature, 30°C).
ρ_f = mass density of air (1.983 × 10 kg/ms at room temperature, 30°C).

Table 2 The drag coefficient for groundnut and soybean (*Source*: Sahay and Singh 1994)

Grain	drag coefficient
Groundnut kernel (RS-1)	0.52-0.64
Soybean (Punjab-1)	0.38-0.62
Soybean (Lee)	0.33-0.51

Table 3 Terminal velocity of different grains (*Source*: Pandey, 1998)

Grains	Terminal velocity m/sec.
Paddy	6–9
Wheat	9–11.5
Barley	8.5–10.5
Oats	8.0–9.0
Small Oats	19.3
Soybean	44.3
Pea	14.25–15.00
Cowpea	12.35–13.29
Sorghum	10.12–10.74
Groundnut Kernel	12.31–13.78
Rye	8.5–10.0
Gram	10.69–13.90
Corn	34.9

Drag coefficient (C) and Drag force (F)

Drag force was calculated based on the formula given by (Mohsenin, 1986; Igwillo, 2017):

$$F_d = C A_p \rho_f \frac{V_t^2}{2}$$

Where,

F_d = drag force (N).

C = Drag coefficient (dimensionless).

A_p = projected area of particle (m^2).

ρf = mass density of air (1.1644 kg/m at room temperature, 30°C).

V_t = Terminal Velocity (m/s).

Conclusion

The filling angle of repose increased linearly as maturity increased. The empting angle of repose increased linearly with an increase in maturity stage. Moisture dependent frictional properties studied namely angle of repose and static coefficient of friction increased with increase in moisture content. Terminal velocity also increased with increase in moisture content, where as rupture force decrease with increase in moisture content.

References

Altuntas, E. and Demirtola, H. (2007). Effect of moisture content on physical properties of some grain legume seeds. New Zealand Journal of Crop and Horticultural Science, 35: 423–433.

Amin, M.N., Hossain, M. A. and Roy, K. C. (2004). Effect of moisture content on some physical properties of lentil seeds. J. Food Eng., 65, 83–87.

Bahnasawy, A.H. (2007). Some Physical and Mechanical Properties of Garlic. International Journal of Food Engineering, 3(6): 1–18.

Davies, R. M. (2009). Some physical properties of groundnut grains. Research Journal of Applied Sciences, Engineering and Technology, 1(2): 464–470.

El Fawal, Y.A., Tawfik, M.A. and El Shal, A.M. (2009). Study on physical and engineering properties for grains of some field crops. Misr Journal of Ag. Engineering, 26(4): 1933–1951.

Gezer, I., Haciseferogullari, H. and Demir F. (2002). Some physical properties of Hacihaliloglu apricot pit and its kernel. Journal of Food Engineering, 56: 49–57.

Gharibzahedi, S.M.T., Mousavi, S.M., Razavi, S.H. and Akavan-Borna, M., (2009). Determination of nutritional and physical properties of sesame seed (*Sesamum Indicum* L.). In: Proceedings of Rural Development Conference. Lithuanian University, pp: 304–309.

Grochowicz, J. (1994). Maszyny do czyszczenia i sortowania nasion. Seed cleaning and sorting machines. Ed. AR, Lublin (in Polish).

Horabik, J. (2001). Charakterystyka własciwosci fizycznych roslinnych materialow sypkich istotnych w procesach skladowania. Acta Agrophysica, 54.

Igwillo, U.C., Eze, P.C. and Eze, C.N. (2017). Selected physical and aerodynamic properties of african breadfruit (Treculia africana) seeds from south eastern Nigeria. Journal of Experimental Research, 5(1): 7–17.

Irtwange, S.V. and Igbeka, J.C. (2002) Selected moisture dependent friction properties of two African yam bean (Sphenostylis stenocarpa) accessions. ASABE, 18(5): 559–565.

Jouki, M. and Khazaei, N. (2012). Some physical properties of rice seed (Oriza sativa). Research Journal of Applied Sciences, Engineering and Technology, 4(13): 1846–1849.

Kabas, O., Yilmaz, E., Ozmerzi, A. and Akinci, I. (2007) Some physical and nutritional properties of cowpea seed (*Vigna simensis* L.). Journal of Food Engineering, 79: 1405–1409.

Kachru, R.P., Gupta, R.K. and Alam, A. (1994). Physico-chemical Constituents and Engineering Properties. Scientific Publishers, Jodhpur, India.

Khodabakhshian, R., Emadi, B., Khojastehpour, M. and Golzarian, M.R. (2017). Physical and frictional properties of pomegranate arils as a function of fruit maturity. International Food Research Journal, 24(3): 1286–1291.

Konak, M., Carman, K. and Aydin, C. (2002). Physical properties of chick pea seeds. Biosystems Engineering, 82: 73–78.

Kram, B.B. (2008). Investigation of the external friction coefficient and the angle of natural repose of cv. Bar and Radames lupine seeds. Inzynieria Rolnicza, 4(102): 423–430.

Lawrowski, Z. (2008). Tribologia. Tarcie, zużywanie i smarowanie. Oficyna Wydawnicza Politechniki Wrocławskiej, Wroclaw.

Mehran, G., Faramarz, K., Taghi, S.M.G., Moayedi, A. and Behnam, K. (2010). Study on Postharvest Physico-Mechanical and Aerodynamic Properties of Mungbean [*Vigna radiate* (L.) Wilczek] Seeds. International Journal of Food Engineering, 6(6): 1–22.

Metcalf, J.R. (1966). Angle of repose and internal friction. International Journal of Rock Mechanics and Mining Science & Geomechanics, 3(2):155–161.

Mohsenin, N.N. (1986). Physical Properties of Plant and Animal Materials. 2nd edition (revised) Gordon and Breach science publishers, New York, USA.

Nalbandi, H., Seiiedlou, S. and Ghassemzadeh, H.R. (2010). Aerodynamic properties of Turgenia latifolia seeds and wheat kernels. Int. Agrophysics. 24: 57–61.

Nimkar, P. M. and Chattopadhyay, P. K. (2001). Some physical properties of green gram. Journal of Agric. Engineering Research, 80: 183–189.

Nosal, S. (2012). Tribologia. Wprowadzenie do zagadnień tarcia, zużywania i smarowania. Wydawnictwo Politechniki Poznańskiej, Poznan.

Nwigbo, S.C., Chinwuko, E.C., Achebe, C.H. and Tagbo, D.A. (2008). Design of breadfruit shelling machine. African Research Review, 2(4): 1–16.

Ozarslan, C. (2002). Physical properties of cotton seed. Biosystem Engineering 83: 169–174.

Pandey, P.H. (1998). Principles and practices of post harvest technology. Kalyani Publishers, New Delhi, Second Edition.

Pandiselvam, R., Kailappan, R., Pragalyaashree, M.M. and Smith, D. (2013). Frictional, Mechanical and Aerodynamic Properties of Onion Seeds. International Journal of Engineering Research & Technology, 2(10): 2647–2657.

Pawanpreet, S. G. and Singh, A.K. (2016). Moisture Dependent Physical and Frictional Properties of Mustard Seeds. International Journal of Engineering Development and Research, 4(4): 464–470.

Razavi, M.A., Mohammad Amini, A., Rafe, A. and Emadzadeh, B. (2007). The physical properties of pistachio nut and its kernel as a function of moisture content and variety. Part III: Frictional properties. Journal of Food Engineering 81: 226–235.

Riyahi, R., Rafiee, S., Dalvand, M.J. and Keyhani, A. (2011). Some physical characteristics of pomegranate, seeds and arios. Journal of Agricultural Technology, 7(6): 1523–1537.

Sacilik, K., Ozturk, R. and Keskin, R. (2003). Some physical properties of hemp seed. Biosystems Engineering. 86, 191–198.

Sahay, K. M. and Singh, K.K. (1994). Unit operations of Agricultural Processing. Vikas Publishing House Private Limited, Delhi-110 014.

Singh, A. K., Kumar, R. and Rai, D. V. (2016). Study on Engineering Properties of Rudraksha (*Elaeocarpus Ganitrus* Roxb.) for Design and Development of Agricultural Processing Units. International Journal of Scientific & Engineering Research,. 7(5): 105–118.

Singh, K.K. and Goswami, T.K. (1996). Physical properties of cumin seed. J. Agric. Eng. Res., 64, (2); 93–98.

Sologubik, C.A., Campan, O.L.A., Pagano, A.M. and Gely, M.C. (2013). Effect of moisture content on some physical properties of barley. Industrial Crops and Products, 43: 762–767.

Suthar, S.H. and Das, S.K. (1996). Some physical properties of karingda [Citrullus lanatus (Thumb) Mansf] seeds. Journal of Agricultural Engineering Research, 65: 15–22.

Tavakoli, M., Tavakoli, H., Rajabipour, A., Ahmadi, H., and Gharib-zahedi, S. M. T. (2009). Moisture-dependent physical properties of barley grains. International Journal of Agric. and Biological Engineering, 2(4): 84–91.

Zareiforoush, H., Komarizadeh, M. H. and Alizadeh, M. R. (2009). Effect of moisture content on some physical properties of paddy grains. Journal of Applied Sciences, Engineering and Technology, 1(13): 132–139.

Zdzislaw, K. (2013). Analysis of frictional properties of cereal seeds. African Journal of Agricultural Research, 8(45): 5611–5621.

6

Rheological Properties of Dough

Namrata A. Giri, and B.K. Sakhale

Abstract

Rheological properties of dough from different flours are very important to know about the kneading and modeling process especially for development of bakery products. Rheological testing is helpful to examine the structural and fundamental properties of dough. The basic principle underlying evaluation of rheological properties are mechanical behavior of the dough, in conjunction with the physical structure of the dough, the molecular structure of the protein continuous phase in the dough and with chemical reactions of functional groups. This chapter presents some of the methods currently used in bakery industry for determining the rheological characteristics of the flour and the dough. The instruments such as Farinograph, Mixograph, Extensigraph, Alveograph and Amylograph, which are used for the measurement of dough rheological properties (due to visco-elastic behavior of dough) were described.

Introduction

Rheology is the study of deformation of material when force is applied. This word is derived from the Greek word 'rheos' which means river, flowing, streaming. So, it also called as "flow science". Rheological studies not only include the flow behaviour of liquids, but also deformation behaviour of solids. Generally, for the determination of rheological properties, material should be subjected to strain over a time and the material parameters such as stiffness, modulus, viscosity, hardness, strength or toughness are determined by considering the subsequent forces or stresses (Dobraszczyk and Morgenstern, 2003). Food rheology is about the flow properties of single food components possess a complex rheological response function. It is the evaluation of effects of processing on food structure and its properties and to find the suitability of food material for baking and product preparation. The composition of processed food and the addition of ingredients to it to have desirable qualities and finished product, require to understand the rheological parameters of single ingredients and their relation to the processing, and their final discernment (Fischer and Windhab, 2011).

Rheology is best tool considered to measure the quantity for the amount of stress required in dough during kneading which is related to the quality of gluten network formed (Bloksma and Bushuk 1988). The basic principles involved in rheological measurements are to get a quantitative description of the material's mechanical properties and information related to the molecular structure and composition of the material along with to characterize and guess the material's performance during processing and for quality control (Dobraszczyk, 2003).

Rheological measurement is also consider as a tool to control and design process and inform about the behavior of dough under given conditions (Scott and Richardson, 1997), and can be used to describe the performance during mechanical operations such as mixing, sheeting (Love *et al.,* 2002; Morgenstern *et al.,* 2002; Binding *et al.,* 2003), proofing (Shah *et al.,* 1999) and baking of dough (Fan *et al.,* 1994). It is applicable to predict the finished product quality such as mixing behaviour, sheeting and baking performance (Dobraszczyk, 2004a; Ross *et al.*, 2004). Due to these benefits of evaluating the product performance and consumer acceptability, rheological instruments and measurement have become essential tools in analytical laboratories (Herh *et al.,* 2005). Herh *et al.,* (2000) reported that, understanding and designing textural, rheological and mechanical properties of different food system is important before food processing. Rheological properties such as behavior of dough are practically very important as it affects directly to the baking performance of flours.

Wheat flour is the main ingredients for the preparation of bakery product and it undergoes several modifications in physical, chemical and biological systems at molecular and micro structural levels (Esselink *et al.,* 2003). When it combines with other ingredients, it's all rheological properties get changed. Thus it is very important to investigate the effects of addition of ingredients and its suitability for preparation of food products.

There are different test methods used to evaluate rheological properties. These techniques are have traditionally been divided into empirical (descriptive) and fundamental (basic) (Bloksma and Bushuk, 1988). Rheological techniques could be categorized into those that investigate dough behavior at constant low temperature to know about the rheological properties of dough during processing stage and second to study the heat involved to know the behavior of dough during baking process (Weipert, 1990). Devices for descriptive empirical measurements of rheological properties, such as the farinograph, extensograph, mixograph, amylograph, and alveograph have been extensively used within the cereals laboratories.

The advantages of use of empirical tests are these are easy to perform and it do not require highly technically trained person to operate and it provides brief information on quality, performance and behaviour of dough such as amount of water required, dough stability etc. (Dobraszczyk and Morgenstern, 2003). They also reported that empirical tests are purely descriptive and dependent on the type of instrument, size and geometry of the test sample and the specific conditions under which the test was performed. Particularly, in these empirical test methods, large deformation force (i.e. large strains) is to be develop and, subsequently, destroy dough structure (Farinograph, Mixolab) or to stretch the dough (Extensograph, Alveograph). Moreover, only one deformation force is used, which results in a single-point measurements.

The present article includes the detailed information about the instruments used for the measurement of rheological characteristics of flour and dough to predict the quality of finished products. The rheological measurement instruments includes; farinograph, extensograph, amylogrph, mixograph and alveograph.

Farinograph

Farinograph (Fig.1) is used to measure and records the mechanical strength of the dough during kneading. It was reported that, the rheological behavior of dough is affected by the type of flour, the quality of the flour and the quality of the other raw materials (Morgenstern *et al.,* 1996). It also helps to find the optimal blends of different flour mixes for the development of quality bakery products. Many researches state that there is close relation between the rheological properties of dough and final quality of product (Osella *et al.,* 2008). Rheological properties of dough also decide the quality of protein which decides the types of flour (Petrofsky and Hoseney, 1995). Various methods have been developed for evaluation of rheological properties of wheat flour dough (Rao and Rao 1993; Peressini, 2001). The addition of protein rich ingredients such as soy, whey protein to wheat flour shows the significant changes of dough farinograph parameters: water absorption, development time and stability of the dough increased (Ammar *et al.,* 2011; Nikolic *et al.,* 2013).

Principle: Farinograph measures the torque needed for mixing dough at a constant speed and temperature (Pomeranz and Meloan 1994). The resistance offered by dough with time and traced on paper in the form of curve. The farinographic parameters are evaluated by using curve.

The resistance to the kneader shaft has increasing when the different ingredients mixed together and water absorption by flour particles, dough formation and development. The consistency is close to the consistency of dough. Then the variation of moment to the shaft, the dough consistency remains constant during dough stability which will be shorter or longer time depends on the flour characteristics. The dough consistency is recorded during kneading on graph termed as farinograma (Fig.2) (Munteanu *et al.*, 2015; Voicu *et al.*, 2012).

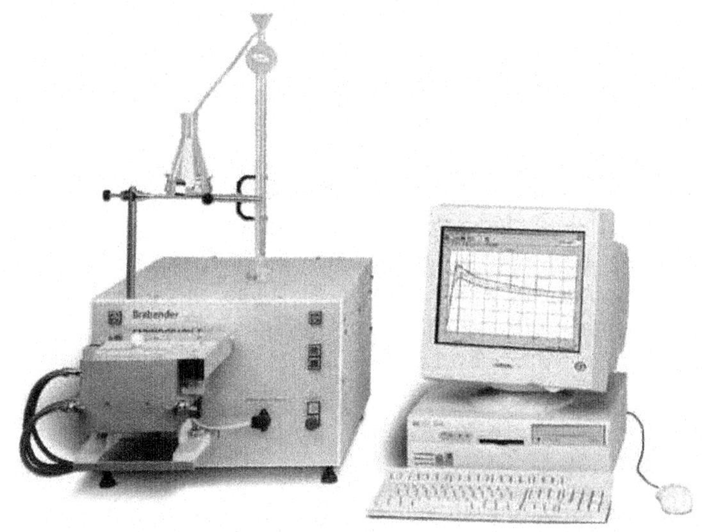

Fig.1 Brabender farinograph (Voicu *et al.*, 2012)

Methodology: The Brabender farinograph (AACC 2000) with electronic record has a mixer type Sigma-300 with a capacity of 300 g flour (450-500 g dough) with dual-casing, through which flows hot distilled water at a temperature of 30 + 1°C, prepared in an outer recirculation bath. The farinograph is fitted with a special strap used to measure the quantity of water introduced into the kneader, respectively for the hydration capacity of the flour. The mixer has two kneading arms Σ-shaped, which rotate in opposite direction. Resistance opposed by the dough at kneader shaft is transferred to a dynamometer and recorded by computer, which translates this information by displaying a graph. The farinograph curve indicates the rheological characteristics of dough on which is evaluated the quality of flour. The software records measurement data, it evaluates them according to standard methods (AACC, ICC) and prints the farinograph curve with data on the properties of flour and dough (Constantin *et al.*, 2011).

The farinograph curve (Fig.2) represents the results such as time of formation or development of the dough, time of stability, the maximum degree of softening of the dough, dough elasticity and flour strength or the farinograph index. On the horizontal axis is the time of kneading from the time of the addition of water, in minutes, such that 10 mm = 1 min (Munteanu *et al.*, 2015).

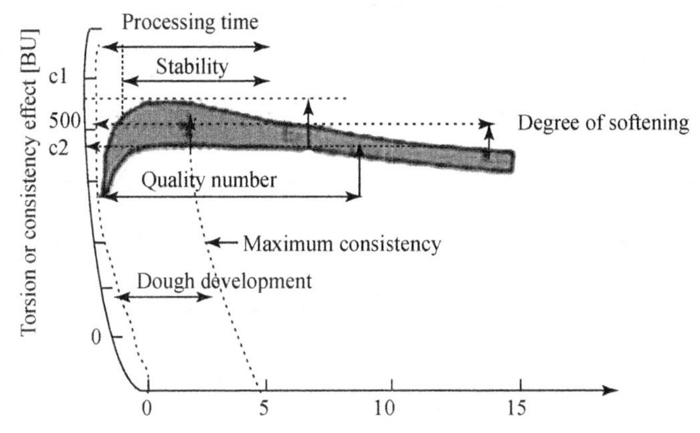

Fig. 2 Farinograph curve (Codină, 2010)

- *Hydration capacity* is the amount of water absorbed by the flour to form dough of standard consistency. It is expressed in ml of water absorbed by 100 g flour. Standard consistency is the consistency of 0.5 kgf m or 500 BU (Brabender units). Typical absorption levels for flour used in bread industry are 58-66% (Panturu and Bîrsan 1999).
- *Time of development (formation) of the dough* is the time required for the formation of gluten, i.e. until the consistency of 500 BU.
- *Dough stability* is the time in which the farinograph curve is maintained on a line of normal consistency.
- *The degree of softening of the dough (the index of tolerance to kneading)* is represented by the difference between the consistency of 500 BU and the consistency reached by the curve after 12 minutes from achieving standard consistency.
- *Volume index of flour (FQN)* is an index for measuring the quality of the flour and is measured on the farinograma, on horizontally (in minutes) from the vertical axis of the consistency of the dough to the point where the center line of the curve meets the horizontal line lowered by 30 FU towards the peak of consistency, multiplied by 10 (Constantin *et al.*, 2011).

Extensograph

The extensograph (Fig.5) is the test to measure the extensibility of dough. The data recorded using extensiograph is dough extensibility and dough resistance to extension.

Principle: The extensigraph determines the resistance and extensibility of dough by measuring the force required to stretch the dough with a hook until it breaks. Extensigraph results include resistance to extension, extensibility, and area under the curve. Resistance to extension is a measure of dough strength. A higher resistance to extension requires more force to stretch the dough. Extensibility indicates the amount of elasticity in the dough and its ability to stretch without breaking.

Methodology: A 300-gram flour sample on a 14% moisture basis is combined with a salt solution and mixed in the farinograph to form dough. After the dough is rested for 5 minutes, it is mixed to maximum consistency (peak time).

Analyses: A 150-gram sample of prepared dough is placed on the extensigraph rounder and shaped into a ball. The ball of dough is removed from the rounder and shaped into a cylinder. The dough cylinder is placed into the extensigraph dough cradle, secured with pins, and rested for 45 minutes in a controlled environment. A hook is drawn through the dough, stretching it downwards until it breaks. The extensigraph records a curve (Fig.3 & 4) on graph paper as the test is run. The same dough is shaped and stretched two more times, at 90 minutes and at 135 minutes.

Results from the Extensigraph Test are useful in determining the gluten strength and bread-making characteristics of flour. (Adapted from Method 54-10, AACC, 2000).

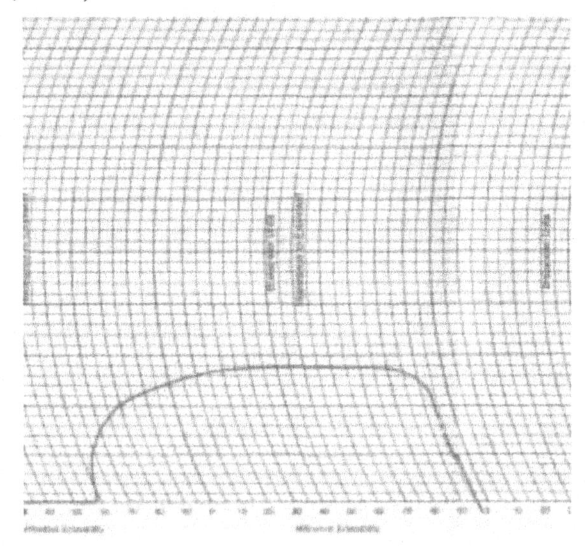

Fig. 3 Extensibility of weak gluten flour

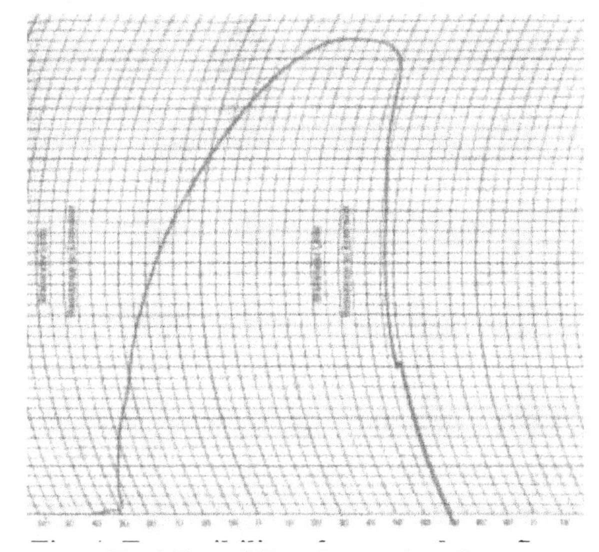

Fig. 4 Extensibility of strong gluten flour

The extensigraph test measures and records the resistance of dough to stretching. Resistance to Extension is the R value and is indicated by the maximum height of the curve. It is expressed in centimeters (cc), Brabender units (BU) or Extensigraph units (EU). Extensibility is the E value and is indicated by the length of the curve. It is expressed in millimeters (mm) or centimeters (cm). R/E Ratio indicates the balance between dough strength (resistance to extension) and the extent to which the dough can be stretched before breaking (extensibility). Area under the curve is a combination of resistance and extensibility. It is expressed in square centimeters (cm^2). Weak gluten flour has a lower resistance to extension (R value) than strong gluten flour.

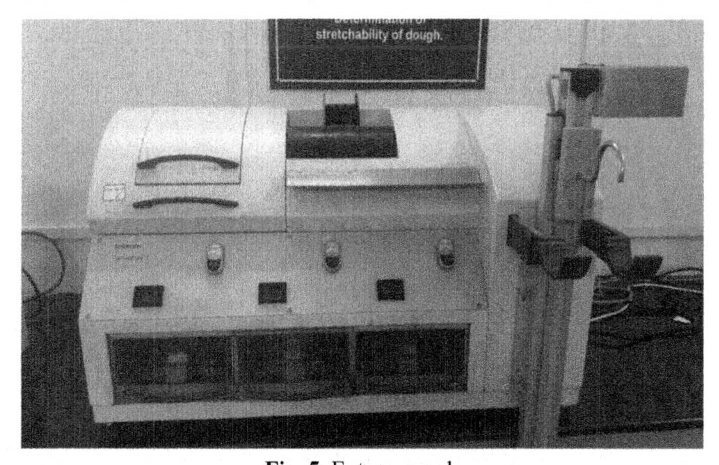

Fig. 5 Extensograph

Amylograph

Amylograph (Fig.8) measures important aspects of starch rheology during pasting (AACC 2000; Pomeranz and Meloan 1994) such as gelatinization temperature and changes in viscosity with temperature and time. These pasting properties of starch help to evaluate crumb characteristics and loaf volume (Kusunose *et al.,* 1999).

Amylograph Characteristics: The different parameters are recoded using amylograph such as:

1. Gelatinization temperature (°C): temperature at which starch starts gelatinizing;
2. Peak viscosity (BU): viscosity in BU at first peak of curve;
3. Temperature at peak (°C): temperature at which peak viscosity is obtained;
4. Viscosity at 95°C (BU): viscosity on attaining 95°C temperature, the relation of this value to peak viscosity reflects the ease of cooking the starch.

Principle: The viscosity of material is calculated by measuring the resistance offered by mixture of flour and water slurry to the stirring action of paddles. The heating action makes starch granules swells and converted into paste. As the slurry becomes paste, then the more resistance to paddles during stirring and has a higher peak viscosity. Generally, a thicker slurry indicates less enzyme activity and makes better products. Amylograph (Figs. 6 & 7) results include peak viscosity.

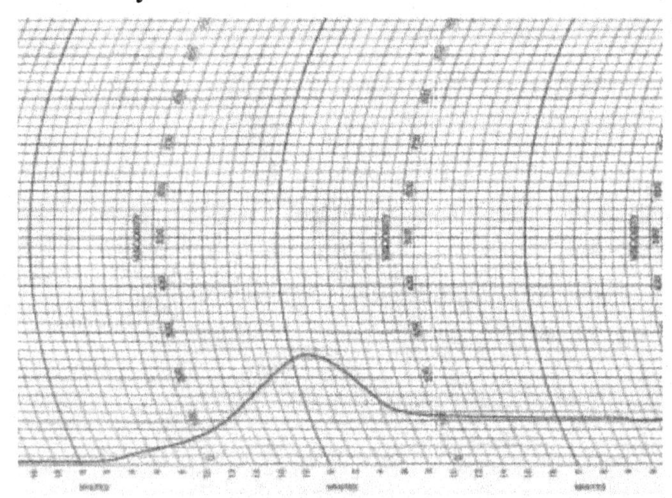

Fig.6 Peak viscosity of Sprouted Wheat Flour (http://www.wheatflourbook.org)

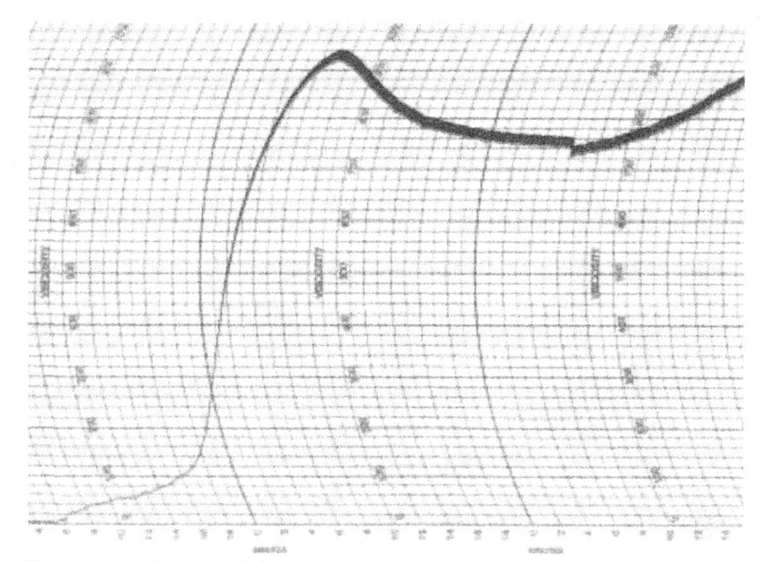

Fig. 7 Peak viscosity of Sound Wheat Flour (http://www.wheatflourbook.org)

Methodology: A 65g of flour mixed with 450ml of distilled water to make slurry. The slurry is continuously stirred and heated in the amylograph, beginning at 30°C and increasing at a constant rate of 1.5°C per minute until the slurry reaches 95°C. The amylograph records the resistance to stirring as a viscosity curve on graph paper.

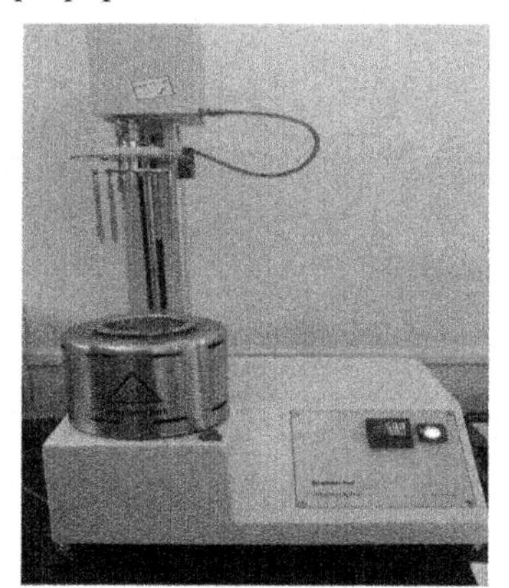

Fig. 8 Amylograph

Mixograph

The mechanism of kneading of dough is recored by mixograph. The four vertical bars of instruments performs the kneading, which is attached to kneading head, rotating the dough into a planetary movement around three fixed straight arms. The kneading force increase as the formation of gluten increases in dough. Maximum capacity for the mixer is 35 g flour.

Principle: The dough and gluten properties of sample is determined by the resistance offered by dough to the mixing action of pins. The mixograph results include water absorption, peak time, and mixing tolerance. The curve obtained shows the gluten strength, optimum dough development time, mixing tolerance (tolerance to over-mixing), and other dough characteristics.

Methodology: In this methos, 10g of flour having 14% humidity placed in the bowl of mixograph. The flour and water mixed together to form dough, and the mixograph records a curve on graph paper. The mixograph curve indicates the strength of gluten, the optimum time of dough development, and other dough characteristics (http://www.wheatflourbook.org). It was reported that, as the mixing operation continued, there is decrease of the mixing curve and the decomposition begins.

Dough behavior is reflected in the tail of the curve, when the mixing continues beyond the peak of mixing this is commonly referred to as mixing tolerance.

Fig. 9 Mixograph (http://www.wheatflourbook.org)

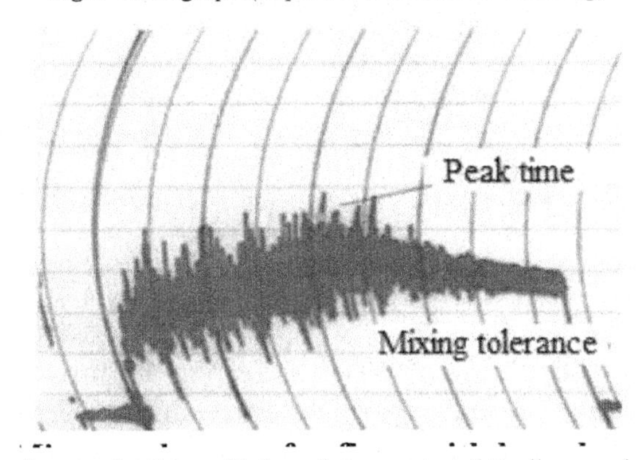

Fig. 10 Mixograph curve for flour with low gluten content (http://www.wheatflourbook.org)

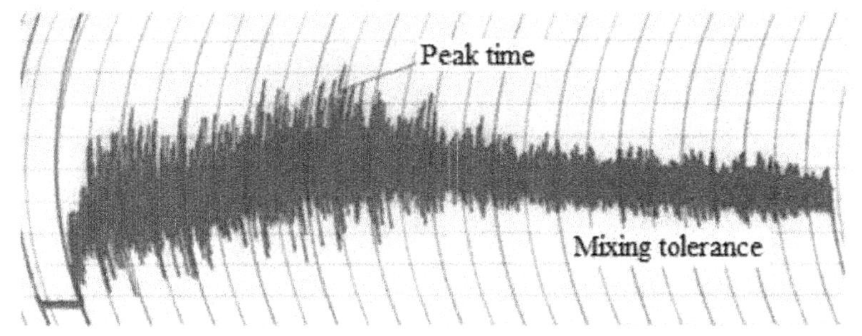

Fig. 11 Mixograph curve for flour with high gluten content (http://www.wheatflourbook.org)

For the development of good quality bakery product, dough should have good tolerance which is the action to stop dough development until the optimal time. Good tolerance would indicate that the bread dough should be elastic after mixing. And failure rate is nothing but the poor stability of dough.

- *Mixing* time is the time required for the optimal development of the dough (the maximum point on the mixing curve or slightly after the peak).
- *Water absorption:* Flour is assessed visually and depends on the protein content, moisture of flour and the environment. Absorption is used to estimate the time of bread baking.
- *Type of flour:* Wheat quality for good bread is largely associated with the quantity and quality of proteins. A high quality of flour will produce good bread in a fairly wide range of percentages of protein, while a variety of relatively poor quality can produce bread of relatively poor quality even when the protein content is high.

The mixograph is drawn depending on protein content / quality and tolerance. Protein content / quality is divided into three categories: low (less than 10%); medium (10% - 12.9%); high (13% or higher) (http://www.wheatflourbook.org).

Alveograph

The alveograph method is based on the tensile strength of a dough sheet subjected to air pressure, which inflates under the form of a bubble that grows until it ruptures (Codină, 2010). The Chopin alveograph (Fig.12) is currently used to determine certain rheological properties such as: tenacity or maximum pressure P, swelling index G, mean abscissa at breaking I and strength energy W (http://www.wheatflourbook.org).

Fig. 12 Chopin alveograph (http://www.wheatflourbook.org)

The registration of the alveographic curve may be a hydraulic gauge, or optionally may be used the Alveolink, an accessory performing the recording of alveographic curves and automatic calculation of the values of P, W, L, G, P/L, Ie.

Principle: The alveograph determines the gluten strength of a dough by measuring the force required to blow and break a bubble of dough.

Methodology: Dough is made from 250 grams of flour mixed with a salt solution; five circular balls of dough are formed, with diameter of 4.5 cm, made by mixing and extrusion, followed by conversion into small disks, which are left in standby for 20 minutes in the alveograph, in a compartment with temperature set to 25 °C; each disk made of dough is tested individually. The alveograph blows air into the disk of dough, which extends into a bubble that eventually breaks; the pressure inside the bubble is recorded as a curve on graph paper.

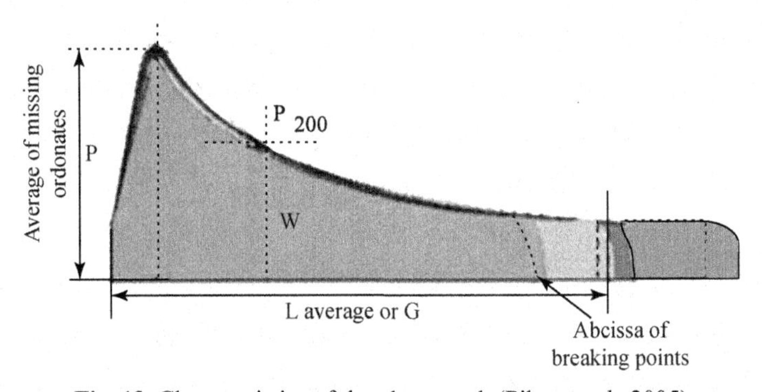

Fig. 13 Characteristics of the alveograph (Pikus *et al.*, 2005)

Values of alveograph (Fig.13) parameters point to the utility properties of flour. Parameter P gives information on flour elasticity and is related to the consistency, plasticity and water absorptiveness of a sample. Parameter L is dough extensibility, depending on elasticity of gluten and the ability of dough to hold gases. Parameter G is equal to square root from the air volume necessary to fill in an air bubble. Parameter W characterizes flour strength and is directly proportional to the area of the alveograph. Parameter P/L reflects the shape of the graph and points to utility properties. Parameter Ie is the index of elasticity and it correlates with dough elasticity (Pikus *et al.,* 2005).

Conclusions

The rheological characteristics of dough are important to assess the quality of different types of flour and in selection of raw materials for the preparation of bakery product. The kneading and shaping of dough plays a crucial role in making finished products. The mechanical and functional properties useful in the bakery food products development are evaluated using farinograph, mixograph, amylograph, alveograph and extensiograph. Both the farinograph and the mixograph evaluate the behavior of flour during kneading, recording characteristics such as: flour hydration capacity, dough development time, dough stability, etc., while the alveograph tests the tensile strength of a sheet of dough subjected to air pressure whereas, extensiograph measures the dough extensibility at different time intervals.

References

AACC. (2000). Approved Methods of the American Association of Cereal Chemists,10th Edition. The Association, St. Paul, MN.

Ammar, A.S., Salem, A.A. and Badr, F.H. (2011). Rheological properties of wheat flour dough as affected by addition of whey and soy proteins. Pakistan Journal of Nutrition,10(4): 302–306.

Binding, D.M., M.A. Couch, K.S. Suyatha and J.F.E. Webster. (2003). Experimental and numerical simulation of dough kneading and filled geometries. J. Food Engg., 58(2): 111–123.

Bloksma, A.H. and W. Bushuk. (1988). World production of wheat and other cereals. In: Pomeranz, Y., (Eds.), Wheat Chemistry and Technology Volume II. American Association of Cereal Chemists, St Paul, Minnesota, USA. pp. 131–218.

Codină, G.G. (2010). Proprietătile reologice ale aluatului din făina de grâu, Seria Inginerie alimentară, Editura AGIR.

Constantin, Gh., Voicu, Gh., Rusanescu, C.O. and Stefan, E.M.(2011). Researches on Rheological Characteristics of Dough of Wheat Flour and their Changes during Storage. Bulletin UASVM Agriculture, 68(2).

Dobraszczyk, B. J. and Morgenstern, M. P. (2003). Rheology and the Bread Making Process. J. Cereal Sci., 38: 229–245.

Dobraszczyk, B.J. (2003). Measuring the Rheological Properties of Dough. In: Bread making Improving Quality. Woodhead Publishing. Cambridge, UK. pp. 375–400.

Dobraszczyk, B.J. (2004a). The physics of baking: rheological and polymer molecular structure function relationships in bread making. J. Non- Newton. Fluid. 124(1-3): 61–69.

Esselink, E., Aalast, H.V., Maliepaard, M., Henderson, T.M.H., Hoekstra, N.L.L. and Duyahoven, J.V. (2003). Impact of industrial dough processing on structure: a rheology, nuclear magnetic resonance, and electron micrograph study. Cereal Chem., 80 (4): 419–423.

Fan, J., Mitchell, J.R. and Blanshard, J.M.V. (1994). A computer simulation of the dynamics of bubble growth and shrinkage during extrudate expansion. J. Food Engg., 23:337-356.

Fischer, P. and Windhab, E.J. (2011). Rheology of food materials. Current Opinion in Colloid and Interface Science, 16:36-40.

Herh, P.K.W., Colo, S.M., Roye, N. and Hedman, K. (2000). Application Note: Rheology of foods: New techniques, capabilities and instruments. Reologica Instruments AB, Sweden.

Herh, P.K.W., Dalwadi, D.H., Roye, N. and Hedman, K. (2005). Flow Control: Rheological Properties of Structural and Pressure-sensitive Adhesives and Their Impact on Product Performance. Reologica Instruments. Sweden.

http://www.wheatflourbook.org

Kusunose, C., Fujii, T. and Matsumoto, H. (1999). Role of starch granules in controlling expansion of dough during baking. Cereal Chem., 76 (6): 920–924.

Love, R.J., Hemar, Y., Morgenstern, M. and McKibbin, R. (2002). Modeling the sheeting of wheat flour dough. Ninth Asian Pacific Confederation of Chemical Engineering Congress APCChE 2002 and 30th Annual Australasian Chemical Engineering Conference CHEMECA 2002, Christchurch, New Zealand.

Morgenstern, M.P., Newberry, M.P. and Holst, S.E. (1996). Extensional properties of dough sheets. Cereal Chemistry, 73(4): 478–482.

Morgenstern, M.P., Wilson, A.J., Ross, M. and Al-Hakkak, F. (2002). The importance of visco-elasticity in sheeting of wheat flour dough. In: Welti-Chanes, J., Barbosa-Canovas, G.V., Aguilera, J.M., Lopez-Leal, L.C., Wesche-Ebeling, P., Lopez-Malo, A. and Palou-Garcia, E. (Eds.). Proceedings of the Eighth International Congress on Engineering and Food Technomic. Puebla, Mexico. pp. 519–521.

Munteanu, M., Voicu, Gh., Stefan, E.M. and Constantin, G.A. (2015). Farinograph characteristics of wheat flour dough and rye flour dough, International Symposium ISB-INMA-TEH, pp. 645–650.

Nikolic, N.C., Stojanovic, J.S. and Stojanovic, G.S. (2013). The effect of some protein rich flours on farinograph properties of the wheat flour. Advenaced technologies, 2(1): 20–25.

Osella, C.A., Robutti, J., Sanchez, H.D., Borras, F. and de la Torre, M.A. (2008). Dough properties related to baking quality using principal component analysis. Ciencia Y Tecnologia Alimentaria, 6(2): 95–100.

Panturu, D. and Bîrsan, I.G. (1999). Manualul Inginerului din Industria Alimentara, Vol. II, Editura Tehnică, Bucuretti.

Peressini, D. (2001). Evaluation methods of dough rheological properties. Tecnica-Molitoria, 52 (4): 337–379.

Petrofsky, K.E. and Hoseney, R.C. (1995). Rheological properties of dough made with starch and gluten from several cereal sources. Cereal Chemistry, 72(I): 53–58.

Pikus, S., Jamroz, J., Olszewska, E. and Wlodarczyk-Stasiak, M. (2005). An attempt to use Saxs method in evaluating different types of wheat flours. Electronic Journal of Polish Agricultural Universities. Vol 8(1): #21

Pomeranz, Y. and Meloan, C.E. (1994). Rheology. Food Analysis: Theory and Practice, 3rd Ed.; Chapman and Hall, Inc.: New York. 449–483.

Rao, G.V. and Rao, P.H. (1993). Methods of determining rheological characteristics of dough: a critical evaluation. J. Food Sci. Technol. 30 (2): 77–87.

Ross, K.A., Pyrak-Nolte, L.J. and Campanella, O.H. (2004). The Use of Ultrasound and Shear Oscillatory Tests to Characterize the Effect of Mixing Time on the Rheological Properties of Dough. Food Res. Int., 37: 567–577.

Scott, G. and Richardson, P. (1997). The application of computational fluid dynamics in the food industry. Trends Food Sci. Technol., 8: 119–124.

Shah, P., G.M. Campbell, C. Dale and A. Rudder. (1999). Modeling bubble growth during proving of bread dough. In: Campbell, G.M., Webb, C., Pandiella, S.S. and Niranjan, K. (Eds.), Bubbles in Food. American Association of Cereal Chemists. St Paul, Minnesota, USA.

Voicu, Gh., Constantin, Gh., Stefan, E.M. and Ipate, G. (2012). Variation of farinographic parameters of doughs obtained from wheat and rye flour mixtures during kneading. U.P.B. Sci. Bull., Series D, 74(2).

Weipert, D. (1990). The Benefits of Basic Rheometry in Studying Dough Rheology. Cereal Chem., 67: 311–317.

7

Dielectric Properties of Foods

Yogesh Kumar, Lochan Singh and Vijay S. Sharanagat

Abstract

The electromagnetic processing instruments such as microwave is very commonly used at house hold level, hence understanding the behavior of food as well as factors affecting its dielectric properties becomes important. Dielectric properties (DEP) play an important in electromagnetic processing of food products and mainly depend upon the moisture content, physical properties, composition, temperature and power applied. Different methods have been used to determine the DEP of food material including parallel plate, coaxial probe, resonance cavity, transmission line and free space method. The selection of these methods is again based on several factors such as type of materials, accuracy of measurement, the scope of the research, frequency range, availability of equipments and resources. Present study thus summarizes the importance of DEP of food, influencing factors and the different measurement techniques.

Introduction

The microwave and radiofrequency (RF) are commonly used techniques in food processing based on electromagnetic energy. These techniques are used for heating, drying, tempering, pasteurization, sterilization, baking, roasting, precooking and reheating of agricultural or food material and are govern by the properties called dielectric properties. The dielectric properties are electrical characteristics which determine the interaction of materials with electrical field. The dielectric properties enable us to understand and model the behavior of food to design the equipment and to explore the new process by the electromagnetic processing. Also, the influence of dielectric properties on electric field enables us to sense some specific characteristics of materials, which further may be linked with the dielectric properties to directly monitor the quality of materials.

In present scenario, the microwave processor is an essential appliance present in most of the kitchens. This is also because the microwave processing is fast and energy saving food processing method compared to the traditional or conventionally used methods. In microwave processing, the energy is transferred directly within materials due to molecular interaction with the applied electromagnetic field. The electromagnetic energy converts into thermal energy when the molecules of the materials directly interact with applied radiation. The ability of microwave to penetrate materials enables volumetric heating. However in contrast to the microwave processing, the conventional heating (Convection/ conduction/radiation)is mainly governed by diffusion of heat from surface to inside of materials accompanied with slow and non-uniform heating in thicker materials. Oppositely, the volumetric and local heating in microwave processing enables the rapid and uniform heating irrespective of the shape and size of the material. The present chapter is thus focused on dielectric properties, methods of measurement, factor affecting and application of microwave in food processing.

Microwave and radio frequency

Electromagnetic waves falling within the 300 MHz to 300 GHz frequency range and the wavelength (λ) range of 1 m to 1 mm are known as microwaves. Whereas, the radiofrequencies are ranged between 10 kHz to about 100 GHz. Federal Communications Commission (FCC) allotted specific number of frequencies which can be used for heating purposes. Frequencies 2450 and 915 MHz are the most commonly used for microwave application at home and industries, respectively. On the other hand, 27 MHz is the most commonly used frequency for radio-frequency (RF) food processing.

The characteristic of electromagnetic waves for the propagation through free space at the velocity of light (c) is defined as:

$$c = \frac{1}{\sqrt{\mu_o \cdot \varepsilon_o}}$$

Where, μ_0 and ε_0 are the permeability (8.854×10^{-12} Fm^{-1}) and permittivity (1.256×10^{-6} Hm^{-1}) of free space, respectively.

However, the electromagnetic waves propagation velocity (v) is affected by the electromagnetic properties of material and it is given as:

$$v = \frac{1}{\sqrt{\mu \cdot \varepsilon}}$$

Where, μ stands for material magnetic permeability and ε stands for electric permittivity.

The frequency of electromagnetic waves (f) is given as below equation if velocity (v) and wavelength of waves (λ) in a material is known.

$$f = \frac{v}{\lambda} \quad \text{or} \quad \omega = 2\pi \cdot \frac{v}{\lambda}$$

Where, ω stands for the angular frequency of the electromagnetic waves.

Principles and mechanism

The absorption of microwave energy in the food materials is mainly governed by ionic interaction and dipole rotation. The ionic interaction is also known as ionic conduction or ionic polarization, in which positively charged ions accelerates in the direction similar to that of the electric field and negatively charged ions in the opposite direction. The nature of electric field is alternating and its direction is frequency dependent. Hence, the direction of the movement of both types of ions changes rapidly. Collision occurs between the moving ions and adjacent particles which set them into more agitated motion. Since the temperature of a material is related to the average kinetic energy of its atoms or molecules, so the agitation of the molecules in this way increases the temperature of the material.

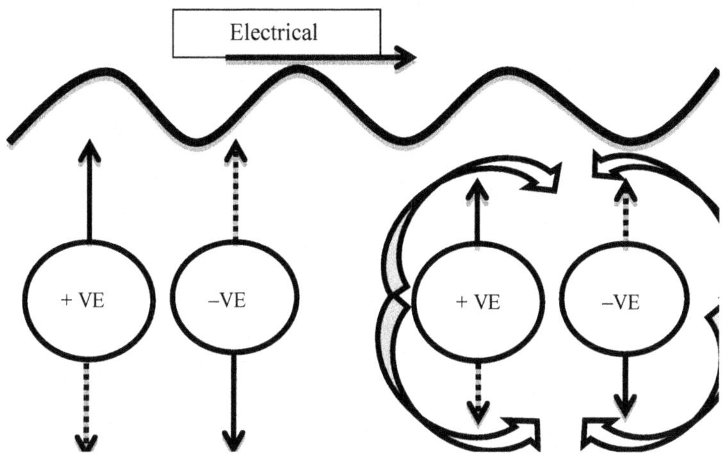

Fig. 1 Mechanism of electromagnetic interaction of agricultural materials. **(A)** Ionic polarisation and **(B)** Dipole rotation

In dipole rotation, the polar molecules orient themselves along the direction of the electric field after their interaction with electric field. They further collide with the other adjacent molecules. Further collisions take place due to alteration in the electrical field's direction. Afterwards, the molecules line up with reverse flow of electric field, thereby generating thermal agitation and heating upon dipole rotation

When a material is exposed to microwave the reflection, transmission and absorption of the energy takes place. The proportions of energy reflected, transmitted and absorbed have been stated in term of the dielectric properties. The most basic electrical property used to describe the interaction is known as complex relative permittivity (ε_r) and expressed as:

$$\varepsilon_r = \varepsilon' - j\varepsilon''$$

Where, real part ε' is dielectric constant, imaginary part ε'' is dielectric loss factor, and $j = \sqrt{-1}$.

Dielectric properties

Dielectric constant (ε') is a material's capacity to store energy when subjected to an electric field. This property modifies the electric field's distribution and phases of waves during their passage within or through the material. Whereas, the loss factor (ε'') describes the capacity to dissipate microwave energy as heat when material is exposed to electric field. The amount converted to thermal energy during the processing of food is proportional to the loss factor value. At radio (1-50 MHz) and microwave frequency (915 and 2450 MHz), the dominant mechanism of energy loss is ionic polarization and dipole rotation. The contribution in energy dissipation by both mechanisms is expressed as:

$$\varepsilon'' = \varepsilon_i'' + \varepsilon_d'' = \varepsilon_i' + \frac{i}{\varepsilon_o \times \omega}$$

Where, ε_i'' and ε_d'' are the contributions of ionic polarization and dipole rotation, respectively; ω stands for angular frequency (Hz); "i" is ionic conductance of material ($S{\cdot}m^{-1}$) and ε_o stands for the vacuum permittivity i.e., 8.854×10^{-12} F m^{-1}.

As electromagnetic waves move through a material, the rate of heat generation per unit volume (i.e., power density per unit volume) decreases which reduces the penetration depth. In a material with high loss factor, the heat generation rate decreases rapidly. The power density per unit volume can be expressed as:

$$P = \frac{\dot{Q}}{V} = \omega \cdot \varepsilon_o \cdot \varepsilon'' \cdot E^2 = 2\pi \cdot f \cdot \varepsilon_o \cdot \varepsilon'' \cdot E^2$$

Where, P stands for power density ($W{\cdot}m^{-3}$); $\dot{Q}$ is heat generation rate (W); V is volume of material (m^3); f is the frequency of wave and E is the root mean square value of electric field strength ($V{\cdot}m^{-1}$).

The incremented temperature of material due to heat generation is also proportional to electric field intensity, treatment time, frequency and the loss factor. The rise in temperature of material due to microwave treatment is given as:

$$\rho \cdot C_p \cdot \frac{dT}{dt} = 55.63 \times 10^{-12} f \cdot \varepsilon'' \cdot E^2$$

Where, ρ and C_p stands for density ($kg{\cdot}m^{-3}$) and specific heat of the material ($J\ kg^{-1}K^{-1}$) respectively and dT/dt = the rate of temperature increase ($K{\cdot}s^{-1}$).

The penetration of electromagnetic energy into the material and attenuation of electric field at the surface is used to estimate the microwave absorbing characteristics of the material used. The distribution of energy inside a material can be estimated by the attenuation factor (α'). The attenuation factor depends on dielectric properties of materials and can be expressed as:

$$\alpha' = \frac{2\pi}{\lambda}\left[\frac{\varepsilon'}{2}\left(\sqrt{1+\tan^2\delta}-1\right)\right]^{1/2}$$

Where loss tangent ($\tan\delta$) is known as the ratio of loss factor (ε'') and dielectric constant (ε').

Dielectric and magnetic properties of the materials critically affect the attenuation of electromagnetic radiation. When the electromagnetic wave attenuation in the material is high, the waves penetrate the material quickly favoring tapering off of the microwave heating. Also, high loss factor exhibits strong attenuation of microwaves demonstrating its potential as a good microwave absorber.

The loss tangent ($\tan\delta$) indicates the electrical field penetration and dissipation of electrical energy as heat. The parameter is known as power penetration depth (d_p) and indicates the distance up to which microwaves will travel inside (penetrate) the material before it reduces by some fraction compared to initial value. The power penetration depth is the depth from surface of sample at which 36.8% ($1/e$) drop inmicrowave power occurs in comparison to its transmitted value. Mathematically depth of power penetration can be derived from Lambert expression as given below:

$$P = P_o e^{(-2\alpha' d)}$$

Where, P is power at depth of penetration d; P_o is incident power at surface; α' is attenuation factor. From definition, $P = \frac{P_o}{e}$ at power penetration depth of d_p.

$$\frac{P_o}{e} = P_o e^{(-2\alpha' d_p)}$$

$$d_p = \frac{1}{2\alpha'}$$

Further, replacing α' and solving it will give mathematical expression for power penetration depth.

$$d_p = \frac{c}{2\pi f\sqrt{2\varepsilon'\left[\sqrt{1+\left(\frac{\varepsilon''}{\varepsilon'}\right)^2}-1\right]}}$$

Where, λ_o is wavelength in free space (m) (for 2.45GHz frequency, λ_o is 0.122 m); f stands for frequency (Hz) and c stands for the speed of light in free space (3×10^8 m/s).The other formula to calculated power penetration depth is:

$$d_p = \frac{\lambda_o \cdot \sqrt{\varepsilon'}}{2\pi\varepsilon''}$$

The basic difference between microwave and radiofrequency heating is in the penetration ability due to difference in wavelength. The less penetration depth of microwave leads to surface overheating with more hot or cold spots and the problem can be countered by using radiofrequency heater.

Dependency of dielectric properties

Dielectric properties of materials depend on both: material's properties and characteristics of microwave. Material's properties such as moisture content, density and composition affect the dielectric properties. In addition to that, the operation condition such as temperature and frequency also affect the dielectric constant and loss factor of any materials.

Effect of moisture: Water is a polar fluid and can easily absorb electromagnetic energy due to dipole rotation. In agricultural or food materials, water exists in either the free form in capillaries or bound form. The stage of water in these materials affects the dielectric properties. The dielectric loss factor remains constant in the bound region up to certain moisture content known as critical moisture. But, when moisture increases beyond critical level, loss factor increases sharply (Fig. 2).

The availability of free water also depends on the interaction of material's components with water. The strong binding forces between carbohydrates or protein and water lead to reduction in free water which reduces the value of loss factor and dielectric constant. Hence, in case of development of new microwaveable food, adjusting the different form of water is very important because of their significant effect on dielectric properties. The pretreatment such as drying reduces the free moisture of the material which further reduces the dielectric properties. The energy conversion also reduces with continuation of drying.

Effect of physical properties: The physical properties of agricultural or food materials such as particle size, bulk density, and homogeneity affects the dielectric properties. Majorly the variation in mass per unit volume affects the loss factor and dielectric constant. This is because dielectric properties are dependent on amount of mass which interacts with the electromagnetic energy. The size of material affects the dielectric constant only. The variation in size modifies the surface characteristics and the proportion of energy transmitted.

The available literatures also suggest that particle size does not affect the loss factor because of similarity between the moisture content and components of the sundry particle size fractions. As a result, there is no change due to heating during size reduction.

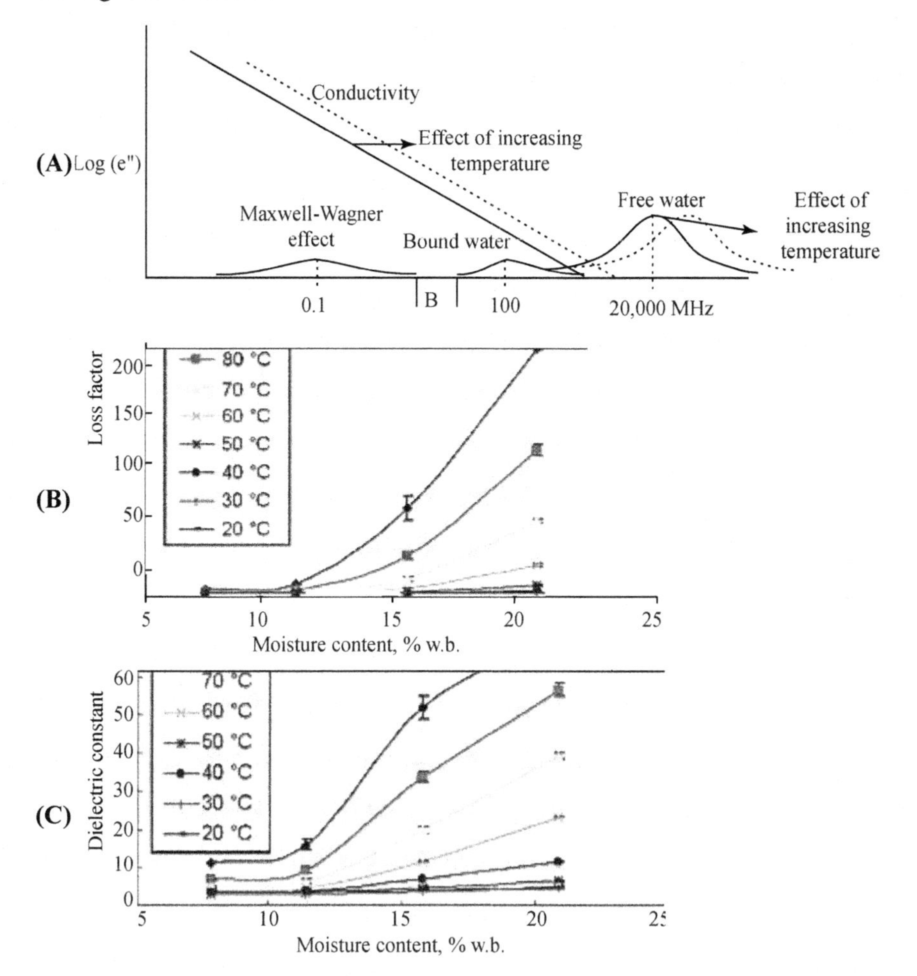

Fig. 2 **(A)** Mechanisms of loss factor for materials as a function of temperature and frequency (Adopted from: Wang *et al.*, 2003). **(B) & (C)** Changes in loss factor and dielectric constant of chickpea flour (27 MHz) due to moisture content and temperature (From: Guo *et al.*, 2008).

Effect of material composition: Water, salt and other minerals influences the dielectric properties of foods but the manner in which they are present in food largely decides the dielectric properties. Free components contribute more to the dielectric properties but their restriction complicates the prediction of the dielectric properties of a composition, based on data for each ingredient. Unlike inorganic components, food-organic constituents are dielectrically inert and may be considered transparent to energy in comparison to ionic fluids or water. In food with lower moisture, water is bound and unaffected by the rapidly changing electromagnetic field. Lower moisture content reduces the loss factor; hence the dehydrated foods have less capacity to convert electromagnetic energy into thermal energy.

Presence of sugars modifies the dielectric behavior of water. While in alcoholic beverages, dielectric properties mainly related to stabilization of hydrogen bonding patterns (Hydroxyl (OH)–Water (H_2O) interaction/ Hydrogen bonding). The presence of fat in food does not affect the dielectric properties significantly. However, increase in fat content reduces the dielectric properties due to the reduction of free water. The presence of protein, free amino acid and polypeptides favors increase in dielectric loss factor. The protein dipole motion depends on their amino acids and pH of medium. The change in protein molecules due to heat is known, as protein denatures also affects the dielectric properties. The disturbance in structure of protein during denaturation results in increase in the asymmetry of the charge distribution which increase the dipole moment and polarization. The presence of ionic compounds significantly affecting the dielectric properties especially the loss factor. The increase in free ions concentration increases the ionic polarization which leads to higher electromagnetic conversion into thermal energy.

Temperature and frequency dependency: The temperature effect on the dielectric properties of biomaterials is quiet complex. The dielectric properties may increment or decrement with increase in temperature and is found dependent on many factors especially the food composition and applied frequencies. The temperature of materials significantly improves the loss factor due to ionic conductance at low frequency electromagnetic energy. However, the high frequency increases the temperature leading to reduction in loss factor because of free water dispersion.

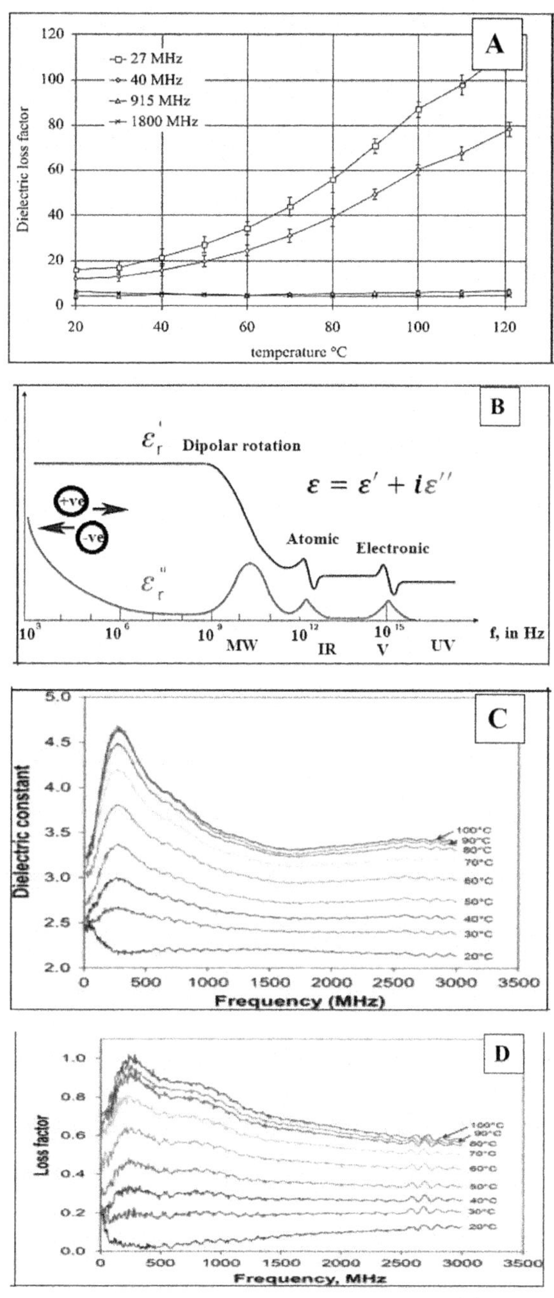

Fig. 3 **(A)** Temperature dependency of dielectric loss factor of cooked macaroni noodles at four frequencies. (From-Source: Wang *et al.*, 2003). **(B)** The Frequency response of different dielectric mechanisms. **(C)** and **(D)** The dielectric properties of egg white powder (8.0% d.b. moisture content and 20 °C to 100 °C temperatures)apropos to frequency (From-Boreddy and Subbiah, 2016)

The dielectric properties of biomaterials considerably vary with the change in frequency of the applied electromagnetic energy. Majorly with the phenomenon of polarization of permanent dipolar molecules which orient with imposed electromagnetic field. Ionic polarization is the main mechanism of energy conversion at lower frequency (< 200 MHz), whereas dielectric dispersion dominates at higher frequency for the most of fruits and vegetables (Fig. 3). However in pure liquids possessing polar molecules as in water and organic liquids like alcohols, dielectric dispersion dominates the frequency characteristics (Table 1).

Table 1 Frequency dependency of liquid water at 20 and 50 °C
(*Source*: Hasted, 1973)

Frequency (GHz)	Dielectric constant		Loss factor	
	20 °C	50 °C	20 °C	50 °C
0.6	80.3	2.75	69.9	1.3
1.7	79.2	7.9	69.7	3.6
3.0	77.4	13.0	68.4	5.8
4.6	74.0	18.8	68.5	9.4
7.7	67.4	28.2	67.2	14.5
9.1	63.0	31.5	65.5	16.5
12.5	53.6	35.5	61.5	21.4
17.4	42.0	37.1	56.3	27.2
26.8	26.5	33.9	44.2	32.0
36.4	17.6	28.8	34.3	32.6

Measurement of dielectric properties

The measurement of dielectric properties is essential for the design of processing equipment and controlling the processing parameters. Dielectric properties of biomaterials can be correlated with other characteristics such as presence of water and density of materials. Hence, the electrical measurement of dielectric properties can be utilized to monitor the other properties of material. There are several measurement techniques available for the measurement of dielectric properties of biomaterials. Also, there are several factors which decide the selection of particular method of measurement. The major factors for selection of particular method are frequency range, type of materials, accuracy of measurement, and the research extent and resource availability for such studies.

Impedance bridges and resonant circuit are traditionally used techniques to study the dielectric properties if frequency for operation is below 200 MHz. While, if operation frequency is more than 200 MHz, transmission line, microwave region and resonant techniques were found useful. We will discuss here five commonly used techniques for dielectric properties measurement.

Parallel plate method: The parallel plate method consists of an arrangement of two parallel metal plates separated from each other by some distance. The parallel metal plates thus form a capacitor. Materials are placed in between the gap the plates. The change in capacitance and dissipation is measured through an impedance analyzer or LCR meter. Loss tangent and dielectric constant can be calculated based on physical dimensions. The main advantage of this method is simple and inexpensive operation with considerably high accuracy. Major method limitation is limited frequency range (< 100 MHz). Additionally, the materials must be able to form smooth flat sheet (< 10 mm thick).

Coaxial probe method: In this method, the dielectric properties are measured by dipping the probe into a liquid or touching it to the flat surface of solid or granular materials. A typical coaxial probe measurement system is made up of a network of coaxial probe, impedance analyzer and appropriate software. The dielectric properties of materials affect the phase and magnitude of reflected signal. The reflected signal is measured by the vector network analyzer and the reflection coefficient is then converted into dielectric properties using software. This method is non-destructive and convenient with easy usage across broad frequency range (200 MHz to 20 GHz). It is found suitable for liquids and semisolids, and no sample preparation is required. The major limitation is its lower accuracy (± 5), low loss resolution and large sample requirement with thickness greater than 1cm.

Transmission line method: In this method, material is placed inside the enclosed transmission line. A typical transmission line technique based measurement system consists of coaxial airline or waveguide section, vector network analyzer, and external computer with suitable software. The reflected and transmitted signals are used for the measurement of dielectric properties. This method is sensitive and accurate compared to the coaxial probe method but it has a narrower working range (<100 MHz) of frequencies. The method is used for the measurement of dielectric properties of all the liquid and solid materials. Major disadvantage is high sample preparation time and the requirement of compact filling of material in the cross section of the transmission without any air gaps.

Resonance cavity method: Resonant cavity method is the most accurate method for measurement of dielectric properties. Resonant cavities are the structures which at certain frequencies show resonance. A very small amount of sample material affects the center frequency and quality factor (Q) of the cavity. Center frequency and quality factor are used to calculate the dielectric constant of the sample. The change in quality factor is used to determine the loss factor. A vector network analyzer is used to find the center frequency and to calculate the quality factor. The major advantages of this method are easy sample preparation, working ability at wide temperature range and less measurement time requirement. Additionally, the method requires a very small amount of sample. On the other hand, it does not provide broadband frequency data and the analysis may be complex. The operation frequency of this method is ranged between 1 MHz to 100 GHz.

Free space method: A free space method does not require the test fixtures to hold the sample. Antennas focus the microwave energy at or through a slab of material. A free-space method based exemplary measurement system consists of a vector network analyzer along with antennas, tunnels, arches and computational unit which measures the reflection and sample transmission. The free space reflection method consist of a transmit horn and a receive horn on the same side of the sample. Whereas, the free space transmission method having two horns on opposite sides of the sample. This method is non-contacting and mostly used for solid materials under hostile environments such as high temperatures. The sample is placed and heated within a furnace with insulated material on windows which gives transparency to microwaves. This method is used to measure large, flat, thin and parallel-faced samples in frequency of microwave region. The calibration of the network analyzer for a free space measurement is major challenge which consumes time. However, recent advances in calibration techniques like Thru-Reflect-Line (TRL) and Thru-Reflect-Match (TRM) are easier than other calibration techniques in free space method.

Applications in food

Dielectric processing has been extensively used for heating, drying, sterilization and extraction based food industries. The main reason to adopt dielectric processing is reduction in processing time and less energy consumption than traditional techniques. As we know now, dielectric heating is volumetric heating techniques, used to instantaneously heat materials due to molecular agitation of polar molecules by dipole rotation and ionic polarization. The efficiency of dielectric heating depends mainly on the material properties; materials with more polar components increase the energy conversion inside. The position of materials inside oven also affects the electromagnetic field distribution due to which non-uniform heating may occur. The uneven heating is intrinsic characteristics and uniform heating is major challenge for dielectric heating, especially in non-homogenous materials. Materials having uneven distribution of polar molecules heat more where concentration of molecules is more and remain cold in region without polar molecules. The dielectric based ovens have turntables to rotate the materials to neutralize the electromagnetic field distribution effects and promote uniform heating. Nowadays, volumetric heating is used in combination with convective, conductive and radiant heating based on quality of desire products. However, dielectric heating leads to chemical changes mainly related to the cooking loss, bioactive components, and antioxidant activity. In some cases, dielectric heating also leads to reduction in anti-nutritional components such as tannins, hemagglutinin activity and phytic acid.

Drying of biomaterials is an important unit operation aimed to reduce the moisture content without affecting the physical nature and chemical composition. Dielectric drying is complex dehydration process based on the volumetric heating and is extensively used due to high drying rate, uniform moisture removal and lower energy consumption in comparison to the traditional drying techniques. During drying, significance portion of energy is required during falling rate drying period which may be significantly reduced by introducing microwave. Traditional drying methods have very long falling rate period and results in the shrinkage and browning. However, dielectric drying increases the temperature inside which generates the vapors inside the materials. The vapors generated are forced to move outside, giving rise to a higher partial pressure gradient that drives vapors to the surface. This reduces the shrinkage and surface hardening of food materials. The constant rate drying period is energy consuming process in dielectric drying due to higher initial moisture. Thus, in order to improve energy efficiency and quality, dielectric drying can be combined with other drying techniques such as hot air drying, vacuum drying, freeze drying and ultrasonic assisted hot air dryer.

Dielectric sterilization is process for pathogens' destruction and enzymes' inactivation to improve the safety and extend the shelf life of biomaterials. There are various mechanisms such as selective heating, cell membrane rupture, electroporation and magnetic field coupling which explain the dielectric sterilization. The selective heating of microbial bodies leads to high rise in the temperature of microbes than surrounding fluid, hence destruction of microbes. In case of the electroporation, the electrical potential difference across the cell membrane results in formation of pores in cells and result in the leakage of cellular material. Finally, magnetic field coupling can destroy the cellular component like protein and DNA coupled in the magnetic field.

Conclusion

The dielectric properties of foods are important in order to understand and model the food behavior apropos to radio-frequency (RF) and microwave processing including heating and cooking. The variation in dielectric properties was observed with the change in moisture content and composition of food as well as the temperature and frequency. The variable characteristics of the biological material and intense use of dielectric equipment in daily life provokes food researchers for extensive research for the development of fast and easy cooking method for food.

References

Boreddy, S.R. and Subbiah, J. (2016). Temperature and moisture dependent dielectric properties of egg white powder. Journal of Food Engineering, 168: 60–67.

Guo, W., Tiwari, G., Tang, J. and Wang, S. (2008). Frequency, moisture and temperature-dependent dielectric properties of chickpea flour. Biosystems Engineering, 101: 217–224.

Hasted, J.B., (1973). Aqueous Dielectrics. Chapman and Hall, London.

http://academy.cba.mit.edu/classes/input_devices/meas.pdf. Agilent Technologies Basics of Measuring the Dielectric Properties of Materials, Application.

http://literature.cdn.keysight.com/litweb/pdf/5989-2589EN.pdf. Key sight Technologies, Basics of Measuring the Dielectric Properties of Materials, Application Note.

Wang, S., Tang, J., Johnson, J.A, Mitcham, E., Hansen, J.D., Hallman, G., Drake, S.R. and Wang, Y. (2003a). Dielectric properties of fruits and insect pests as related to radio frequency and microwave treatments. Biosystem Engineering, 85 (2): 201–212.

Wang, Y., Wig, T., Tang, J. and Hallberg, L. (2003). Dielectric properties of food relevant to RF and microwave pasteurization and sterilization. Journal of Food Engineering, 57: 257–268.

8

Thermal Properties of Foods

Vaishali, Samsher, Harsh P. Sharma, Vipul Chaudhary
Sunil and Ankur M. Arya

Abstract

Thermal properties of food are its ability to conduct, store and lose heat. These properties are inherent in food processing and preservation practices. Thermal properties are important for modelling processes, engineering design of processing equipment, sterilization and aseptic processing. Important thermal properties are thermal conductivity, specific heat and thermal diffusivity. These properties depend on chemical composition, water fraction and temperature.

Introduction

Thermal properties are involved in almost every food processing operation. It is essential to have knowledge of thermal properties of foods which is important in analysis and design of various food processes and food process equipment involved in heat transport. Thermal properties of food are greatly depending on composition, temperature and density of food. Thermal properties required for heat transfer calculation include specific heat, enthalpy, thermal conductivity and thermal diffusivity. Thermal properties of foods are related to heat transfer control in specified foods and can be classified as thermodynamic properties (enthalpy and entropy) and heat transport properties (thermal conductivity and thermal diffusivity). Thermo physical properties not only include thermodynamic and heat transport properties, but also other physical properties involved in the transfer of heat, such as freeze and boiling point, mass, density, porosity, and viscosity. These properties play an important role in the design and prediction of heat transfer operations during the handling, processing, canning, storing, and distribution of foods.

Thermal conductivity

The thermal conductivity of a material is defined as a measure of its ability to conduct heat. It also defined as the rate of heat flow through unit thickness of material per unit area normal to the direction of heat flow at per unit time for unit temperature differences. It has a unit of W/m K in the SI system. A solid may be comprised of free electrons and atoms bound in a periodic arrangement called a lattice. Thermal energy is transported through the molecules as a result of two effects: lattice waves and free electrons. These two effects are additive. In pure metals, heat conduction is based mainly on the flow of free electrons and the effect of lattice vibrations is negligible. In alloys and non-metallic solids, which have few free electrons, heat conduction from molecule to molecule is due to lattice vibrations. Therefore, metals have higher thermal conductivities than alloys and non-metallic solids.

Thermal conductivity can be expressed as

$$Q = KAdT$$

Where

Q = amount of heat flow, kcal

A = area, m^2

dT = temperature difference in the direction of heat flow, °C

K = thermal conductivity, kcal/h m °C

Measurement of Thermal Conductivity: Measurement of thermal conductivity can be done by either steady-state or transient-state methods. The advantages of steady-state methods are the simplicity in the mathematical processing of the results, the ease of control of the experimental conditions, and often quite high precision in the results. However, a long time is required for temperature equilibration. The moisture migration and the necessity to prevent heat losses to the environment during this long measurement time are the disadvantages of steady-state methods (Ohlsson, 1983). In addition, these methods require definite geometry of the sample and relatively large sample size. On the other hand, the transient methods are faster and more versatile than the steady-state methods and are preferable for extensive experimental measurements. Transient methods are preferred over steady-state methods because of the short experimental duration and minimization of moisture migration problems. Steady-state methods are longitudinal heat flow, radial heat flow, heat of vaporization, heat flux, and differential scanning calorimeter methods. The most important non steady methods are the thermal conductivity probe method, transient hot wire method, modified Fitch method, point heat source method, and comparative method (Rahman, 1995).

Thermal conductivity of cheese was measured using the line heat source technique described by Murakami *et al.,* (1996). The details of the basic theory and mathematical derivation behind the use of the line heat source probe have been discussed previously by Hopper and Lepper 1950; Nix *et al.,* 1967 and Murakami *et al.,* 1996 gave a comprehensive discussion about the design of thermal conductivity probe and its limitations. A line heat source probe was designed and fabricated for this study. The probe was consisted of a 3.8 cm long stainless steel needle tubing (dia: 0.635 mm), enclosing an insulated constantan heater wire (diameter: 0.076 mm), and an insulated E-type thermocouple wire (dia: 0.051 mm), which measured temperature at the mid-point along the needle.

The thermal conductivity measurement was conducted by inserting the probe into the middle of a cheese block, and then heat was supplied by a heater for two minutes, and its time-temperature data were monitored every two seconds. After the few seconds of heating, the temperature started to rise linearly from about 21 to 25°C with natural logarithm of time and the slope of this portion was determined. The sample's thermal conductivity (k) was determined by dividing the slope (DTemperature = Dln(time)) with a probe factor (G). The probe factor was determined by running the experiment with a calibration sample (water gelled with 0.8% agar) with known thermal conductivity value of 0.626 W/m°C. The experimental error of this probe factor was about 6%. The probe was further tested by measuring other materials of known k-values, such as NaCl brine (30% w/v and 15% w/v; gelled with 0.5% agar) and Glycerol (100%) and their k-values were within 6% of their published values.

The thermal conductivity of Cheddar cheese tested in this study ranged from 0.354 to 0.481 W/m°C with high quality data as measured by their coefficient of variation (CV) of less than 5%. Table 1 shows the thermal properties values of Cheddar cheese with its chemical composition. Thermal conductivity of Cheddar cheese increased with moisture content, decreased with fat content and increased with protein content (Marschoun *et al.,* 2001). Thermal conductivity increases with particle size and thickness or bulk density. Increasing the bulk density means increasing the number of particles in a constant volume thus decreasing the pore volume which leads to increased heat conduction ability of the sample (John *et al.,* 2014). Other researchers also found thermal conductivity to be increasing by temperature and as the thickness (equivalent of bulk density) Taiwo *et al.,* (1996); Aviara and Haque (2001) and Bart-Plange *et al.,* (2009) observed an increase in thermal conductivity with temperature for sheanut kernel and maize and cowpea respectively. Kurozawa *et al.,* (2008) found thermal conductivity to increase from 0.57 to 0.61 W/m°C with temperature in the range of 25 to 45 °C for cashew apple. Both thermal properties were found to increase with increasing temperature. A similar behaviour was observed by Hu and Mallikarjunan (2005) for oysters, Telis-Romero *et al.,* (1998) for Brazilian orange juice; Singh and Goswami (2000) for cumin seed. Thermal conductivity of food depended on the structure and chemical composition of the sample and it increased with increasing water content for all food products at temperature above freezing. These were obtained in this study for the thermal conductivity of soups by Olayemi and Rahman (2013).

Table 1 Composition of Cheddar Cheese with Their Thermal Properties Values (Marschoun *et al.*, 2001)

S. No.	Moisture Content (%)	Fat (%)	Protein (%)	Thermal Conductivity (k, W/m°C)	Thermal Diffusivity (m^2/s)	Heat Capacity (KJ/Kg °C)
1.	33.65	34.94	26.09	0.354	1.18	N/A
2.	34.80	35.15	26.09	0.356	1.09	N/A
3.	44.23	23.40	28.36	0.391	1.17	2.902
4.	47.43	17.20	29.75	0.423	1.17	2.761
5.	50.00	17.00	30.28	0.432	1.10	2.923
6.	54.92	9.40	32.55	0.472	1.34	2.970
7.	37.94	32.25	26.72	0.388	1.12	2.451
8.	45.58	19.35	28.02	0.428	1.16	2.601
9.	50.15	10.20	31.04	0.465	1.17	2.804
10.	45.44	21.65	30.28	0.425	1.21	2.640
11.	41.15	24.00	28.47	0.422	1.23	2.796
12	39.42	25.80	28.06	0.410	1.18	2.609
13	47.09	17.16	30.12	0.448	1.25	2.940
14.	37.00	N/A	N/A	0.369	1.14	2.639
15.	36.69	32.25	N/A	0.353	1.24	2.340
16.	33.82	33.25	N/A	0.338	1.16	2.429
17.	36.20	34.38	N/A	0.366	1.26	2.414
18.	40.54	33.25	N/A	0.381	1.27	2.524
19.	36.02	34.00	25.06	0.359	1.21	2.444
20.	38.00	29.80	N/A	0.356	1.20	2.481
21.	47.20	26.60	N/A	0.403	1.23	2.660

Specific heat

Specific heat is the amount of heat required to increase the temperature of a unit mass of the substance by unit degree. Therefore, its unit is J/kg K in the SI system. The specific heat depends on the nature of the process of heat addition in terms of either a constant pressure process or a constant volume process. The specific heats of foodstuffs depend very much on their composition. Knowing the specific heat of each component of a mixture is usually sufficient to predict the specific heat of the mixture. Specific heat is measured by methods of mixture, guarded plate, comparison calorimeter, adiabatic agricultural calorimeter, differential scanning calorimeter (DSC).

During the course of this study, three DSC (differential scanning calorimeter) were used to measure the heat capacity of Cheddar cheese. They were manufactured by Netzsch (Model DSC-200), by Perkin Elmer (Model DSC-7) and by TA Instruments (Model 2920 Modulated DSC).

The equipment was calibrated according to the manufacturer's specification. After calibration, the base line was established by running the program with no sample present. Weighed empty aluminum sample pans were placed in both sample and reference holders and scanned at a programmed heating rate of 5°C/min over a selected temperature interval (0-100°C). In order to give the initial and final transient behaviour time to disappear, the temperature program was started at 10°C and ended at 110°C. The procedure was repeated with a known mass of sapphire standard and samples of cheese [25-30mg]. DSC thermoforms typically resulted in seemingly random, very noisy observations between 0 and 50°C. Christenson *et al.*, (1989) observed a similar erratic behaviour and attributed it to endothermic peak caused by melting fat. For this reason the average heat capacity values from 60 to 90°C was considered as the value of the measurement. Three samples were used for each experiment and the range of experimental errors was 5-15%. Due to the presence of moisture, hermetically sealed sample pans were used. A pan sealer was used to seal the sample pan to prevent moisture losses occurring during heating. After completion of each run the sample pan was reweighed to ensure there was no loss of sample mass during the run.

The heat capacity of Cheddar cheese tested in this study ranged from 2.444 to 3.096 kJ/kg C with CV ranged from 5-15%. The heat capacity of Cheddar cheese increased with moisture and protein content but decreased with fat content. Table1 shows a highly significant correlation between heat capacity of Cheddar cheese and moisture, fat, and protein content.

The specific heat increased linearly with increasing moisture content and the linear relationship was established. The specific heat capacity obtained ranged from 3.3315 ± 0.00 to 3.8505 ± 0.00 kJ/kg/K. These values are slightly below the specific heat capacity of water. It has been proved that there is direct correlation between the specific heat capacity and moisture content of food product. This implies that as the moisture content increases, the specific heat capacity increases (Lewicki, 2004). This was noticed in the soup samples, with Ewedu having the highest moisture content and specific heat capacity as shown in Table 2 and 3.

Table 2 Proximate composition of some selected nigerian soups (Olayemi and Rahman, 2013)

Soups	MC%	Protein, %	Fat %	Ash %	CF %	CH O %	E(k/cal/10 g)
Ogbono	68.70 ± 0.14	18.70 ± 0.42	6.12 ± 0.11	4.55 ± 0.21	1.04 ± 0.60	1.90 ± 0.01	133.08 ± 0.60
Ewedu	88.60 ± 0.14	6.00 ± 0.01	1.05 ± 0.05	1.81 ± 0.01	1.47 ± 0.02	1.05 ± 0.04	34.27 ± 1.89
Ila	77.25 ± 0.35	15.94 ± 0.08	2.125 ± 0.04	1.90 ± 0.14	1.15 ± 0.07	1.475 ± 0.11	87.61 ± 3.31
Kuka	78.54 ± 0.06	8.80 ± 0.14	2.29 ± 0.01	2.09 ± 0.01	0.875 ± 0.02	7.415 ± 0.08	85.64 ± 0.17

MC = Moisture Content(% wb); CF = Crude Fibre(% wb);
Protein Content (% wb); E = Energy, k/cal/10g, CHO = Corbohydrate's

Table 3 Thermal properties of some selected nigerian soups (Olayemi and Rahman, 2013)

Soups	CP (kJ/kg/K)	K (W/m/K)	D (m^2/s)
Ogbono	3.3315 ± 0.00	0.4470 ± 0.00	1.22
Ewedu	3.8505 ± 0.00	0.5295 ± 0.00	1.36
Ila	3.5535 ± 0.01	0.4825 ± 0.00	1.28
Kuka	3.5860 ± 0.00	0.4935 ± 0.00	1.30

C_p = Specific Heat Capacity, KJ/ Kg/K; K = Thermal Conductivity, W/m/K;
D = Thermal Diffusivity, m/s^2

Thermal diffusivity

Thermal diffusivity (α) is a physical property associated with transient heat flow. It is a derived property. The unit of thermal diffusivity is m^2/s in the SI system. It measures the ability of a material to conduct thermal energy relative to its ability to store thermal energy. Materials with large thermal diffusivity will respond quickly to changes in their thermal environment while materials of small thermal diffusivity will respond more slowly, taking longer to reach a new equilibrium condition.

Thermal diffusivity can be calculated by Indirect Prediction Method and Direct Measurement Methods. Thermal diffusivity can be calculated indirectly from the measured thermal conductivity, density, and specific heat. This approach needs considerable time and different instrumentation. Direct measurement of thermal diffusivity is usually determined from the solution of one dimensional unsteady-state heat transfer equation. The most commonly used thermal diffusivity measurement methods are the temperature history method, thermal conductivity probe method, and Dickerson method (Rahman, 1995).

Thermal Diffusivity Measurement: This study measured thermal diffusivity of Cheddar cheese, as well as comparing two different methods for obtaining this property. The first one was using an apparatus described by Dickerson, (1975). It consisted of an agitated water bath at 60°C in which a steel cylinder (20 cm in length and 4 cm in diameter), insulated with rubber corks on both ends, containing the cheese sample was immersed. Type T thermocouple (diameter: 0.051 mm) was soldered to the outside surface of the tube monitoring the surface temperature of the sample. The centre temperature of the sample was measured by inserting a stainless steel needle tubing (OD: 0.889 mm; L: 7.4 cm), containing a thin type T thermocouple, through the centre of the upper rubber cork. After a period of equilibration between the sample temperature and the water temperature, the heater was turned on for at least one hour or until the sample centre temperature increases at the constant, water-bath heating rate (about 0.67C/min). The measurements were repeated three times for each experiment. The second method for determining thermal diffusivity values was using time-temperature history method. The underlying principle and the mathematical theory behind the unsteady state heat transfer for infinite cylinders can be found in Charm (1978). He has solved the unsteady state heat transfer for an infinite cylinder at a uniform initial temperature and exposed to a constant-temperature environment as follows

$$\frac{T_a - T}{T_a - T_i} = \frac{2}{R}\sum_{n=1}^{\infty} \frac{J_0\left(B_n \frac{r}{R}\right)\left[\exp\left(\frac{-B_n^2 k\theta}{\rho C_p R^2}\right)\right]}{\left(1+\frac{k^2 B_n^2}{h^2 R^2}\right)\frac{B_n}{R}\left[J_1(B_n)\right]}$$

where, T_a = ambient temperature (°C), T_i = initial temperature (°C), T = temperature at time θ(°C), R = radius of the cylinder (m), J_o (X) = zero order of first kind Bessel function of X, B_n = root of m $J_o(B_n) = B_n J_1(B_n)$, where $m = hR/k$, r = distance from the centre (m), k = thermal conductivity (W/m °C), θ = time (s), r = density (kg/m3), C_p = heat capacity (J/kg °C), h = surface heat transfer coefficient (W/m^2 °C), $J_1(X)$ = first order of first kind Bessel function of X.

Large measurement errors resulting from the use of the Dickerson method to determine thermal diffusivity suggest this method is not appropriate to measure thermal diffusivity of Cheddar cheese. To verify this limitation, an experiment was conducted to measure thermal diffusivity of distilled water gelled with 0.3% agar. The value was found to be 1.556107 m^2/s which were in good agreement with results reported by Dickerson and Read (1975); Rizvi *et al.*, (1980).

Aviara and Haque (2001), Taiwo *et al.*, (1996) and Bitra *et al.*, (2010) all tested their materials in granular form for their various samples and were able to establish a linear relationship between thermal diffusivity and moisture content. The thermal diffusivity quantifies a material`s ability to conduct heat relative to its ability to store heat. The results obtained indicated that thermal diffusivity increased with increase in moisture content.

Conclusion

The thermal properties of foods are important in the design of food storage and refrigeration equipment as well as in the estimation of process times for refrigerating, freezing, heating, or drying of foods. The thermal properties of foods are strongly dependent upon chemical composition and temperature. Thermal conductivity increases with thickness and moisture content. Specific heat and thermal diffusivity increases with moisture content.

References

Aviara, N.A. and Haque, M.A. (2001). Moisture dependence of thermal properties of sheanut kernel. Journal of Food Engineering, 47: 109–113.

Bart-Plange, A., Asare, V. and Addo, A. (2009). Thermal conductivity of maize and cowpea. Journal of Engineering and Technology, 2-3: 6–11.

Bitra, V.S., Banu, S., Ramkrishna, P., Narender, G. and Womac, A.R. (2010). Moisture dependent thermal properties of peanut pods, kernels and shell. Journal of Food Engineering, 106: 503–512.

Charm, S. (1978). In: Fundamentals of Food Engineering. AVI: Westport, CT, 156–188.

Christenson, M.E., Tong, C.H. and Lund, D.B (1989). Physical Properties of Baked Products as Function of Moisture and Temperature. Journal of Food Processing and Preservation, 13: 201.

Dickerson, R.W. and Read, R.B. (1975). Thermal Diffusivity of Meats. Trans ASHRAE, 81:356.

Hooper, F.C., Lepper, F.R. (1950). Transient Heat Flow Apparatus for the Determination of Thermal Conductivities. ASHRAE Trans, 56: 309–324

Hu, X. and Mallikarjunan, P. (2005). Thermal and dielectric properties of shucked oysters. LWT- Food Science and Technology, 38(2): 489–494.

John, I., Olugbenga, O.S. and Kayode, Q. (2014). Determination of Thermal Conductivity and Thermal Diffusivity of Three Varieties of Melon. Journal of Emerging Trends in Engineering and Applied Sciences, 5(2): 123–128

Kurozawa, L. W., Park, K. J. and Azonbel, P. M. (2008). Thermal conductivity and thermal diffusivity of papaya (*Carica papaya*, L.) and Cashew apple (*Anacardium occidentale* L.). Brazilian Journal of Food Technology, 11:78–85.

Lewicki, P.P. (2004). Water as the determinant of food engineering properties: A review. Journal of Food Engineering, 61: 483–495.

Marschoun, L. T., Muthukumarappan, K. and Gunasekaran, S. (2001). Thermal properties of cheddar cheese: experimental and modelling. International Journal of Food Properties, 4 (3): 383–403

Murakami, E.G., Sweat, V.E., Sastry, S.K., Kolbe, E., Hayakawa, K. and Datta, A. (1996). Recommended Design Parameters for Thermal Conductivity Probes for Non-frozen Food Materials. Journal of Food Engineering, 27 (2): 109–123.

Nix, G.H., Lowery, G.W., Vachon, R.I., Tanger, G.E. (1967). Direct Determination of Thermal Diffusivity and Conductivity with a Refined Line Source Technique. Progress in Aeronautics and Astronautics: Thermo-physics of Spacecraft and Planetary Bodies, 20: 865–878.

Ohlsson, T. (1983). The measurement of thermal properties. In R. Jowitt (Ed.), Physical Properties of Foods. London: Applied Science.

Olayemi, R.A. and Rahman, A. (2013). Thermal properties of some selected nigerian soups. Agricultural Sciences, 4: 96–99.

Rahman, M.S. (1995). Food Properties Handbook. New York: CRC Press.

Rizvi, S.S.H., Blaisdell, J.L. and Harper, W.J (1980). Thermal Diffusivity of Model Meat Analog Systems. Journal of Food Science, 45: 1727–1731.

Singh, K.K. and Goswami, T. K. (2000). Thermal properties of cumin seed. Journal of Food Engineering, 45(4): 181–187.

Taiwo, K.A., Akanbi, C.T., and Ajibola, O.O. (1996). Thermal properties of ground and hydrated cowpea. Journal of Food Engineering, 29: 249–256.

Telis-Romero, J., Telis, V.R.N., Gabas, A.L., and Yamashita, F. (1998). Thermophysical properties of brazilian orange juice as affected by temperature and water content. Journal of Food Engineering, 38: 27.

9

Optical Properties: An Introduction and Application for Quality Assessment of Fruits and Vegetables

Krishna Kumar Patel, Yogesh Kumar and Yashwant Kumar Patel

Abstract

Quality assessment of fruits and vegetable before marketing and processing is an essential task. Various destructive and non-destructive methods have been investigated to achieve this goal. Among the several prevalent noninvasive methods, optical properties measurement methods also promotes for nondestructive quality inspection of fruits and vegetables. The optical properties of biological tissue are, therefore, very importance important for getting sample information and data quantification of the inside of the material. The main goal of this chapter is, thus, to introduce fundamental concept of optical properties, its principle and application in fruits and vegetables. The discussion of electromagnetic radiation for better understanding of the nature of light has also been explored.

Introduction

India ranks second after china in the production of both fruits and vegetables. India, the important fruit and vegetable producing country in the world, support more than 17 % of the population with only 2.4 % land share. Fruits and vegetables account for nearly 90% of the total horticulture production in the country (Neeraj *et al.*, 2017). Production of vegetables India is estimated at 181 mt in 2017-18, about 1% higher than the year before, while that of fruits is estimated at 95 mt, 2% higher than the previous year (GOI, 2018). India accounts for 16 per cent of world production of vegetables and 11 per cent of the world's fruit production. Presently horticulture contributes 28 per cent of agricultural GDP (Gross domestic product) but only 2 per cent of horticulture produce is processed (Neeraj *et al.*, 2017). In spite of such remarkable production of fruits and vegetables, only 0.4 per cent is exported and 22 per cent is lost or get wasted in market chain (Amarasinghe *et al.*, 2007) because of lack of rapid, non-destructive and precision sorting methods for quality and safety assurance. A non-invasive optical technique using absorption and scattering coefficients has been applied in engineering research for the evaluation of crops and agricultural produce. The optical properties (absorption and scattering coefficients) of agricultural produce are normally measured from the physical state governing interaction of light with the material. The biochemical factors could be vital in quality changes including metabolic and hormone control during the development of maturity stages and the absorption of chemical constituents. Likewise, the material can be categorized according to the optical parameters (absorption and scattering) that are particularly related to the material. This method ultimately relies on the measurement of absorption and scattering properties measurement of various types of turbid agricultural produce (Qin and Lu, 2008). The measurement of optical properties, thus, can provide abundant information about the physico-chemical characteristics of tissues and can help to achieve the goal of nondestructive evaluation of food and agricultural produce. The interest of researchers in the application of optical characterization of food and agricultural produces for nondestructive quality assessment has gained momentum (Qin and Lu, 2007, 2008).

The main objective of this chapter is therefore, to discuss the basics of optical properties, electromagnetic spectrum, components and application optical properties for evaluation of quality of fruits and vegetables.

Electromagnetic spectrum

The range of frequencies (the spectrum) of electromagnetic (Fig. 1) radiation and their respective wavelengths and photon energies are known as electromagnetic spectrum. It is made up of light of many different wavelengths. Most wavelengths are invisible to us and our eyes can only detect small portion of spectrum between 400 to 700 nanometers (nm), called as visible light. Electromagnetic spectrum is divided into broad categories of gamma rays, X-rays, ultraviolet, optical, infrared, sub-millimeter (microwaves), and radio. These categories are arranged in decreasing order of energy and increasing order of wavelength. Gamma ray photons have the most energy and smallest wavelengths of any part of the EM spectrum, in the range of 100,000 eV and above, with corresponding wavelengths of 0.01 nm or less. Gamma rays are produced by the hottest objects in the universe, including neutron stars, pulsars, and supernova explosions. In addition, these rays can also be created by nuclear explosions and the majority of gamma rays, however, generated in space are blocked by the Earth's atmosphere. This is a good thing as gamma rays are biologically hazardous.

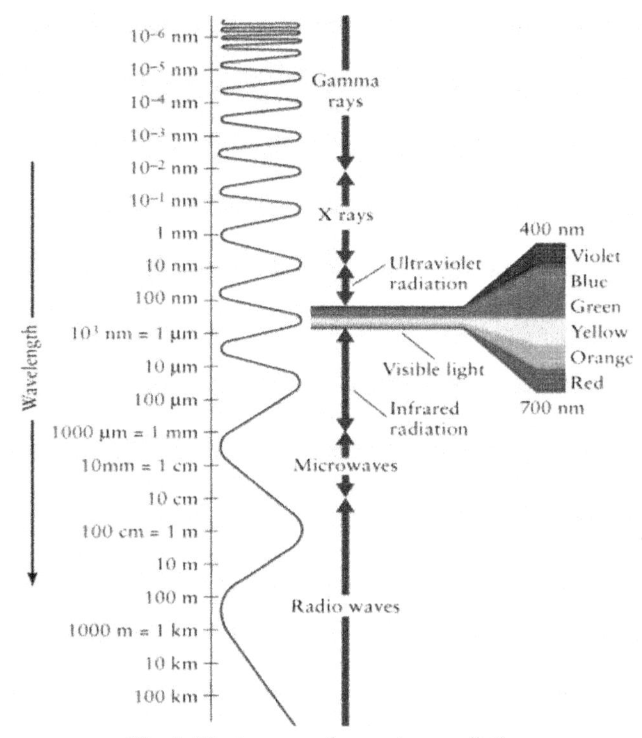

Fig. 1 Electromagnetic spectrum radiation

X-Rays lie in wavelength range from 0.01 – 10 nm. These rays are primarily generated from by super-heated gas from exploding stars and quasars. X-rays are able to pass through many different types of materials and are commonly used for medical imaging and for inspecting cargo and luggage. Similar to gamma rays, the Earth's atmosphere blocks x-ray radiation. Ultraviolet (UV) light has wavelengths of 10 – 310 nm. The Sun is a source of ultraviolet energy.

The UV portion of the spectrum is subdivided into UV-A, UV-B, and UV-C. UV-C rays are the most harmful and are almost completely absorbed by our atmosphere. UV-B rays are the harmful rays that cause sunburn. Although UV waves are invisible to the human eye, some insects, such as bumblebees, can see them. The small part of the electromagnetic spectrum that human eyes can detect is visible light. Visible light when travels through a prism/ diffraction grating, it becomes separated into the colors of the rainbow. Each color of visible light has a different wavelength. Light at the lower end of the visible spectrum, having a longer wavelength (about 740 nm) is seen as red; light in the middle of the spectrum is seen as green; and light at the upper end of the spectrum having wavelength (about 380 nm) is seen as violet. All wavelengths of light move at the same speed but different wavelengths have different energy. The wavelength red light is longest but it is the least energetic while violet light has the shortest wavelength and is the most energetic. The visible portion of the spectrum is used extensively in remote sensing and is the energy that is recorded using photography. The wavelength of infrared portion of the spectrum ranges from approximately 700 nm to 100000 nm (Near Infrared, NIR: 700 – 1300nm, Shortwave Infrared, SWIR: 1300 – 3000 nm and the Far or Thermal Infrared 3000 – 100000 nm).

Infrared radiation is used extensively in remote sensing. Objects reflect, transmit, and absorb the Sun's near-infrared and shortwave radiation in unique ways and this can used to observe the health of vegetation, soil composition and moisture content. The region from 8 to 15 μm is referred to as thermal infrared since these wavelengths are best for studying the long wave thermal energy radiating from the Earth. Furthermore, microwaves, wavelengths range from 1mm to 1m, are essentially high frequency radio waves. Different wavelengths or bands of microwaves are used for different applications. Mid-wavelength microwaves can penetrate haze, light rain and snow, clouds, and smoke are beneficial for satellite communication and studying the Earth from space. Radar technology sends pulses of microwave energy and senses the energy reflected back. Radio waves have the longest wavelengths in the electromagnetic spectrum with wavelengths ranging from approximately 1 mm to several hundred meters. Radio waves are used to transmit a variety of data such as wireless networking, television, amateur radio, etc. all use radio waves while the use of radio frequencies is usually regulated by governments.

Visible, infrared, and ultraviolet laser beams, thus, can provide continuous inspection through scanning of product size, symmetry, damage, irregular shape, fill level, and label placement by adding automatic software in connection with mechanical devices. For example, during conveying of pre-fried potato chips, optical devices detect any with defects (for example, black spots), and automatically deploy an air nozzle to deflect their path from the conveyor belt.

Optical Properties

The interaction between light and an object creates characteristics of the physical phenomena (i.e., changes in the light path and/or light quality) which depend mainly on the nature of light and on the composition and structure of the material. The latter is responsible for the optical properties of said material. Some chemical and physical material properties modify light, as it goes through a transparent object or reflects off of an opaque object (Duran and Calvo, 2003). In fruits and vegetables (turbid biological materials) light interaction involved both absorption and scattering phenomenon. The complex nature of light propagation in biological tissue is because of simultaneous occurrence of specular reflection, absorption and scattering. In a strongly scattering material, photons often undergo multiple scattering before being absorbed or exiting from the material. Whole light, only except 4 to 5% of incident light on turbid biological tissue that is reflected in the form of specular reflection, passes through the surface and interacts with the internal components (Mireei, 2010).

Light absorption is primarily related to chemical constituents of the material (e.g., sugar), whereas light scattering is influenced by structural/physical characteristics (density, particle size, and cellular structures). These two fundamental optical properties are characterized by the absorption coefficient (μ_a) and reduced scattering coefficient (μ_s) (Tuchin, 2000). The classification of objects into transparent, semi-transparent and turbid generally depends on their optical properties and reemerging light caries significant information of biological tissues (Zidouk and Styles, 2011; Hu *et al.*, 2015). The optical parameters of fruits and vegetables can be determined by measurement of the specular reflection, light absorption, and scattering. Some tissue's optical parameters and related terminologies have, therefore, been described below.

Absorption

Light absorption is the process by which the incident radiant flux is converted to another form of energy, usually heat, or into photons with a much lower frequency (e.g., fluorescence) (Bass *et al.*, 2001).

Absorption coefficient (μ_a): Absorption coefficient (mm^{-1}) is an optical parameter that measures how quickly the incident light loses intensity per incremental path length due to light absorption alone (Jacques, 2013; Tuchin, 2007). It is defined as:

$$\mu_a = -\frac{1}{T}\frac{dT}{dL}$$

Where, T (dimensionless) is the transmitted or surviving fraction of the incident light after an incremental path-length dL (mm). The fractional change dT/T per dL yields an exponential decrease in the intensity of the light as a function of increasing path-length L.

Anisotropy factor (g): It is defined as the average of the cosine of the scattering angle, as opposed to isotropy, which implies identical properties in all directions (Tuchin, 2007) and describes extremely total forward scattering (Mie scattering at large particles), isotropic scattering (Rayleigh), and highly backward scattering, when g approaches 1, 0, and -1, respectively. The anisotropy factor is a measure of the amount of photons retained in the forward direction after a single scattering event. Further, it is also a measure of the asymmetry of the single scattering pattern, which means a difference in tissue physical or mechanical properties when measured along different axes. The anisotropy factor (g) can be mathematically expressed as:

$$g \equiv \langle \cos\theta \rangle = \int_0^\pi p(\theta)\cos\theta \cdot 2\pi \sin\theta d\theta$$

Where, $p(\theta) = \frac{1}{4\pi}\frac{1-g^2}{(1+g^2-2g\cos\theta)^{3/2}}$ is the scattering phase function and θ is the angle of light scattered from the incoming direction.

Albedo (a): The albedo (a) is the probability of survival for a photon incident on a small-volume element. It is equal to the ratio of the scattering and total attenuation, ranging from zero for a completely absorbing medium to unity for a completely scattering medium (Tuchin, 2007). The mathematical expression of albedo is given below:

$$a = \frac{\mu_s}{\mu_s + \mu_a}$$

Where, μ_s and μ_a are the scattering and absorption coefficients, respectively.

Attenuation coefficient (μ)**:** The attenuation coefficient, which is equal to the sum of the scattering and absorption coefficients, can be determined by measuring the collimated transmission of a light beam by optically thin slabs of tissue (Zijp and ten Bosch, 1998). By application of Beer's law we can calculate:

$$\mu = \mu_a + \mu_s = -\frac{\ln T_c}{d}$$

with μ the attenuation coefficient in mm^{-1}, μ_s the scattering coefficient in mm^{-1}, μ_a the absorption coefficient in mm^{-1}, T_c the collimated transmission and d the thickness in mm.

Effective penetration depth (δ): Effective penetration (cm) is depth from the surface where the light intensity is equal to the light compensation point (Wang *et al.*, 2015).

$$\delta = \frac{1}{\mu_{eff}} = \frac{1}{\sqrt{3\mu_a\left[\mu_a + \mu_s(1-g)\right]}}$$

Effective attenuation coefficient (μ_{eff})**:** Effective attenuation coefficient (EAC, cm^{-1}) is proportional to the geometric mean of absorption and reduced scattering coefficients (Antonio *et al.*, 2017) and can be expressed as:

$$\mu_{eff} = \sqrt{3\mu_a\left[\mu_a + \mu_s(1-g)\right]} = \sqrt{3\mu_a(\mu_a + \mu_s')} \approx \sqrt{3\mu_a\mu_s'}$$

Where, μ_a and μ_s' are the absorption and reduced scattering coefficients, respectively

Mean free path (*mfp*): The mean free path (cm) defines the mean path length that a photon has to travel before it encounters either absorption or a scattering event (Lorenzo, 2012). The mean free path expressed as:

$$mfp = \frac{1}{\mu_a + \mu_s(1-g)}$$

Reduced scattering coefficient (μ_s'): The reduced scattering coefficient (cm^{-1}) is a lumped property that is derived from the scattering coefficient and anisotropy factor and can be expressed as:

$$\mu_s' = \mu_s(1-g)$$

Refractive index (n): By measurement of the specular reflection, light absorption, and scattering, the tissue optical parameters can be determined. The refractive index is an optical parameter that combines the effects of the different molecules or scattering particles (Bashkatov *et al.*, 2011)

Scattering coefficient (μ_s): The scattering coefficient (cm^{-1}) measures how quickly the incident light loses intensity per incremental path-length due to light scattering alone (Jacques, 2013; Tuchin, 2007). It is defined as:

$$\mu_s = -\ln \frac{T_c}{L}$$

Where, T_c is a collimated transmittance measurement, and L is tissue thickness. The measurement must be made through a thin tissue sample (one *mfp* or less); otherwise, multiple scattering becomes an issue.

Scattering

Light scattering is a physical interaction occurring in the tissue due to localized non-uniformities in the medium through which the light passes, causing the light to deviate from a straight trajectory. Scattering has many forms, including Rayleigh, Mie and Raman scattering (Bass *et al.*, 2001).

Reflection

The light incidenting at an opaque material will be reflected in two different ways. A small part of the incident light bounces on the surface and comes out at the same angle as the incident beam and with the same quality or spectral distribution of energy. This is called specular reflection (S) usually occurs on a polished/gloss smooth surface and it is equal to the incident angle of the light. Since it only interacts with the surface, specular reflection is largely determined by the surface properties and carries little information about the internal tissue (Hu *et al.*, 2015).

The main part of the incident light will undergo diffused reflection (D), i.e., it will travel somewhat through the outer layers of the material, where it is partly absorbed, more for some energy frequencies than others, and consequently, it is altered in its spectral quality. This diffused light will come out in practically all directions and is responsible for what is called the color of opaque objects (Duran and Calvo, 2003).

Color

Color is an optical property responsible for visual appearance and it is no doubt the most important optical property of nearly all materials particularly foods. Visual appearance also includes other characteristics of foods, like shape, size, surface and flesh structure, gloss, translucency, and defects. Among these, only gloss and translucency are important optical properties. In addition, there are other optical properties, like fluorescence and phosphorescence, but of no relevance to foods. It was Newton who first introduced the scientific knowledge of color by launching his theory of colors in 1666.

Methods for optical properties measurement

Optical properties are generally obtained by using solutions of the radiative transport equation. These solutions are either exact or approximate and correspond to the direct or indirect methods. Measurement of optical parameters, thus, can be made using either direct or indirect methods (Wilson *et al.*, 1987).

Direct methods: Direct methods are based on some fundamental concepts and rules. These methods place stringent constraints on the sample to match the assumptions made for the exact solution, such as the single-scattering phase function for thin samples or the effective light penetration depth for slabs. Direct methods have the advantages of being model independent and simple in their analytic expressions for data processing, but the samples tested should be thin (<*mfp*) in order for single light scattering to dominate without complicating the measurement (Wilson *et al.,* 1987). For example, the direct method used thin samples in which multiple scattering could be ignored (Hu *et al.*, 2015).

Indirect methods: Indirect methods relax the sample constraints but require approximations that are often invalid for tissue samples (e.g., nearly isotropic scattering or no internal reflection at the boundaries). Optical measurements with an indirect method can be done on bulk or intact tissue and these methods can be subdivided into iterative and non-iterative. The theory used in indirect methods usually falls into one of three categories: Beer's law, Kubelka-Munk, or the diffusion approximation. Methods based on the diffusion approximation or a similar approximation, for instance, uniform radiances over the forward and backward hemispheres; tend to be more accurate than the Kubelka-Munk method (which is limited in their accuracy). Similarly, Beer's law neglects scattering and is inappropriate for thick scattering materials. Techniques using the diffusion approximation include pulsed photo-thermal radiometry, time-resolved spectroscopy (TR), radial reflectance spectroscopy, weak localization, and an iterative technique that uses reflection and transmission measurements. These methods are popular because of ease application, place relatively minor constraints on the type of sample and are amenable to analytic manipulation (Prahl *et al.* 1993). Although indirect methods are appealing, numerous limitations still need to be overcome. For example, the mathematical models and algorithms can be quite complicated because the irradiation and detection geometries must be accounted for. In general, the simpler models, such as dual-flux and diffusion approximation, are accurate only over a limited range of optical properties and geometries. The lack of analytic inverse solutions may also present a practical difficulty. Analytic inverse solutions are necessary to iterate between the required fundamental properties and the derived data to obtain a better-fit solution. Some major characteristics of different direct and indirect methods are listed in Table 1 (Cheong *et al.*, 1990; Tuchin, 2007; Hu *et al.*, 2015).

Table 1 Direct and indirect method s for optical properties measurement (Hu *et al.*, 2015)

Methods	Nature	Class	Parameters*		Reference
			Measured	Estimated	
Lambert-Beer law	Direct	-	R, T	μ_a, μ_s, g	Pfefer *et al.*(2003)
Kubelka-Munk	Indirect	Non-iterative	R, T	A_{KM}, S_{KM}	Gaonkar *et al.* (2014)
Time-resolved	Indirect	Iterative	$R(r, t)$	μ_a, μ_a, g	Vanoli *et al.* (2014)
Frequency Domain	Indirect	Iterative	$R(r, f)$	μ_a, μ_s, g	Cletus *et al.* (2010)
Spatially resolved	Indirect	Iterative	$R(r)$	μ_a, μ_s, g	Qin and Lu (2008); Trong *et al.* (2014)
Integrating sphere	Indirect	Iterative	R, T	μ_a, μ_s, g	Zhang *et al.* (2012)
Optical coherence tomography	Indirect	Iterative	$i(z)$	g, μ_s, μ_t	Verboven *et al.* (2013)

*R = diffuse reflectance, T = diffuse transmittance or collimated transmittance, r = distance between source and detector, t = time, f = modulation frequency, i = heterodyne signal current, and z = depth.

Application of optical parameters in quality assessment

Quality assessment is an essential operation in fruits and vegetables processing industry not only to satisfy the consumer but also to monitor safety and to develop a checklist for inspection to obtain high-quality produce. Optical properties measurement in this case could provide accurate results and it is feasible for any specific application. The measurement of optical properties has been tested on several agricultural produce and its application is increasing day-by-day in the form of nondestructive tool.

Lu (2004b) developed a hyperspectral imaging-based SR method that is faster in measurement, easier to use, and less expensive, compared to the TR and FD methods, for measuring the optical properties of fruits and vegetables. They have applied spatially resolved techniques for measuring the optical properties of horticultural and food products (Qin and Lu, 2007). Further, an optical property measuring prototype, called the Optical Property Analyzer (OPA), which integrated hyperspectral imaging with the optimized inverse algorithm, was also developed to automatically measure the optical properties of fruits and vegetables (Cen, 2011). Several studies have been reported by the researchers using these of developed techniques for the measurement of the optical properties, specially absorption and scattering properties of citrus fruit (Lorente *et al.*, 2015), apples (Beers *et al.*, 2014), cucumber (Cen *et al.*, 2014), bananas (Hashim *et al.*, 2012, 2013), peach (Cen *et al.*, 2012), apple (Lu *et al.*, 2010), kiwifruit (Baranyai and Zude, 2009), and squash of tomato, pear, kiwifruit, plum and zucchini (Qin and Lu, 2008) and juices of fruit and vegetable (Qin and Lu, 2007). The measured optical properties were also used for predicting quality attributes and chemical constituents (firmness, soluble solids content, fat content, etc.).

Table 2 Application of optical properties for quality assessment of fruits and vegetables

Fruits and Vegetables	Quality parameters/objects	Optical parameters	References
Apple	Fruit development observation	μ_a, ′	Seifert *et al.*, (2015)
	Internal browning assessment	μ_a	Vanoli *et al.*, (2014)
	Chilling injury	D	Baranyai *et al.*, (2011)
	For selection of apples of distinctive quality	μ_a, μ_s'	Rizzolo *et al.*, (2010)
	Chilling injury	μ_a, μ_s, g	Baranyai *et al.*, (2009)

Fruits and Vegetables	Quality parameters/objects	Optical parameters	References
Citrus	For monitoring changes in optical properties	μ_a, μ_s'	Cubeddu *et al.*, (2001a)
	Assessment of physico-chemical properties	μ_a, '	Cubeddu *et al.*, (2001b)
	Decay	$\mu_a, \mu_s', R_r(r)$	Lorentz *et al.*, (2015)
	Decay	Gaussian-Lorentzian parameters	Lorentz *et al.*, (2013)
Cucumber	For measurement of optical properties, firmness/quality	$\mu_a, \mu_s', R_r(r)$	Qin and Lu (2008)
Kiwifruit	To study light propagation	μ_a, μ_s, g	Baranyai and Zude (2009)
	Optical properties measurement, ripeness, firmness/quality	$\mu_a, \mu_s', R_r(r)$	Qin and Lu (2008)
	Assessment of physico-chemical properties	μ_a, μ_s'	Cubeddu *et al.*, (2001b)
Nectarine	Maturity assessment	μ_a	Tijskens *et al.*, (2007)
Onion	Diseased onion detection	μ_a, μ_s', g	Wang *et al.*, (2014)
	For estimating optical properties	μ_a, μ_s', g	Wang and Li (2013)
	To study light propagation		Wang and Li (2013)
Peach	For assessment of chemical and physical properties	μ_a, μ_s'	Cubeddu *et al.*, (2001b)
	For measurement optical properties	μ_a, '	Qin and Lu (2008)
	For maturity assessment	μ_a, μ_s'	Cen *et al.*, (2012)
Pear	For maturity assessment	μ_a, μ_s'	Nicolai *et al.*, (2008)
	For detection of brown heart	μ_a, μ_s'	Zerbini *et al.*, (2002)
Tomato	Ripeness grading	μ_a, μ_s'	Zhu *et al.*, 2015)
	Ripeness, firmness	$\mu_a, \mu_s', R_r(r)$	Qin and Lu (2008)
	Leaf minor damage	Reflectance spectral parameters	Xu *et al.*, (2007)
	For assessment of chemical and physical properties	μ_a, μ_s'	Cubeddu *et al.*, (2001b)

μ_a: Absorption coefficient; m$_s$: Scattering coefficient ; μ_s' : Reduced scattering coefficient; $R_r(r)$: Relative refractive index; *g*: Anisotropy factor and *D*: Diffusion Coefficient

Furthermore, Loeb and Barton (2003) acquired OCT (Optical coherence tomography) images of kiwifruit, orange flesh, orange peel, lettuce leaves, and intact cranberries. They reported that OCT could be a valuable tool for non-destructive, morphological imaging of botanical subjects.

Further, hyperspectral imaging system has been used by Erkinbaev *et al.* (2011) to investigate optical properties of an apple. In this they studied on two types of apple samples (soft and hard apples) and reported that there were three types of dimensional hyper cubes (one consisting of spectral axis and two spatial axes) were obtained based on selected wavelength. The sample achieved higher absorption values in the range of 650–670 nm as compared to the absorption values in the range of 500-600 nm. In an another study by Beers *et al.* (2014), potential of a hyperspectral scatter imaging system has been discussed for evaluating ripeness level of apples in the range of 550-1000 nm. They have reported that the application of hyperspectral imaging is beneficial for using in optical sensor devices and automated harvesting systems. Similarly, Verboven *et al.* (2013) have reported that OCT (Optical coherence tomography) offered 2D and 3D images with high resolution and a large field-of-view for visualizing the structural differences in apple peel and for measuring the structural changes during storage. The optical parameters, therefore, can be vital for classification and grading of fruits and vegetables/agricultural produce (Table 2).

Conclusion

There is rapid introduction of optical parameter's measuring techniques in the field of quality assessment of agricultural produce such as apple, citrus, cucumber, kiwifruits, pear, plum, peach, onion, tomato, etc. Quality assessment of fruits and vegetables by optical mainly based on the structural complexity, diversity and inhomogeneity. Detection of bruise/defects, decay and chilling injury and assessment of maturity, firmness, ripeness, internal browning, etc are few common application of this technique. In spite of above achievements and future prospects, the modern research on optical properties measurement techniques is not advanced as expected. Further, most of the optical instrument initially developed for biomedical applications and they are still too expensive. In addition, the determination of optical parameters of fruits and vegetables is totally numerical. As errors can only be obtained from the quadrature point of the light propagation medium, the implementation of numerical analysis is impractical. Conclusively, it therefore, can be said that there is a need of further work to facilitate optimum usage of this technique not only for assessing of fruits and vegetables but also in processing industries of other agricultural produces.

References

Amarasinghe, U.A., Shah, T., Turral, H. and Anand, B., 2007. India's water futures to 2025–2050: Business as Usual Scenario and Deviations. IWMI Research Report 123. Colombo, Sri Lanka: International Water Management Institute.

Antonio, M.C., Maclin, E.L., Low, K.A., Fantini, S., Fabiani, M. and Gratton, G. (2017). Low-resolution mapping of the effective attenuation coefficient of the human head: a multidistance approach applied to high-density optical recordings. Neurophotonics. 4(2): 021103

Baranyai, L. and Zude, M. (2009). Analysis of laser light propagation in kiwifruit using backscattering imaging and Monte Carlo simulation. Comput. Elec. Agric., 69(1), 33–39.

Bashkatov, A.N., Genina, E.A. and Tuchin, V.V. (2011). Optical properties of skin, subcutaneous, and muscle tissues: A review. J. Innov. Optical Health Sci, 4(1), 9–38.

Bass, M., Van Stryland, E.W., Williams, D.R. and Wolfe, W.L. (2001). Handbook of Optics: Vol. III. Visual Optics and Vision (2nd ed.). New York, N.Y.: McGraw-Hill.

Beers, R.V., Gutiérrez, L.L., Schenk, A., Nicolaï, B., Kayacan, E. and et al. (2014). Optical measurement techniques for the ripeness determination of Braeburn apples. In Beers, R. V., Gutiérrez, L.L. and Schenk, A. (Eds). Proceedings of the Int. Conf. Agric Eng, p. 1-7. Switzerland: Agricultural Engineering Zurich.

Cen, H. (2011). Hyperspectral imaging-based spatially resolved technique for accurate measurement of the optical properties of horticultural products. PhD diss. East Lansing, Mich.: Michigan State University.

Cen, H., Lu, R., Ariana, D.P., and Mendoza, F. (2014). Hyperspectral imaging-based classification and wavebands selection for internal defect detection of pickling cucumbers. Food Bioproc. Tech., 7(6), 1689–1700.

Cen, H., Lu, R., Mendoza, F. and Ariana, D. (2012). Assessing multiple quality attributes of peaches using optical absorption and scattering properties. Trans. ASABE, 55(2), 647-657.

Cheong, W.-F., Prahl, S. A., and Welch, A. J. (1990). A review of the optical properties of biological tissues. IEEE J. Quantum Elec., 26(12), 2166–2185.

Cletus, B., Künnemeyer, R., Martinsen, P. and McGlone, V.A. (2010). Temperature-dependent optical properties of Intralipid measured with frequency-domain photon-migration spectroscopy. J. Biomed. Optics, 15(1), 017003.

Cubeddu, R., D'Andrea, C., Pifferi, A., Taroni, P., Torricelli, A. and et al. (2001a). Time-resolved reflectance spectroscopy applied to the nondestructive monitoring of the internal optical properties in apples. Appl. Spectroscopy, 55(10), 1368–1374.

Cubeddu, R., D'Andrea, C., Pifferi, A., Taroni, P., Torricelli, A. and et al. (2001b). Nondestructive quantification of chemical and physical properties of fruits by time-resolved reflectance spectroscopy in the wavelength range 650–1000 nm. Appl.Optics, 40(4), 538–543

Duran L and Calvo C (2003) Optical Properties of Foods. Food Engineering, book edited by Barbosa-Canovas GV, Vol. 1, 1–9.

Erkinbaev, C., Tsuta, M., Nguyen, N., Trong, D., Verboven, P. and et al. (2011). Hyperspectral scatter imaging for contactless food quality evaluation. In Erkinbaev, C., Tsuta, M. and Nguyen, N. (Eds). Proceedings of the 11th Int. Congress on Eng Food, p. 1–5. Greece: ICEF Publications.

Gaonkar, H.A., Kumar, D., Ramasubramaniam, R. and Roy, A. (2014). Decoupling scattering and absorption of turbid samples using a simple empirical relation between coefficients of the Kubelka-Munk and radiative transfer theories. Appl. Optics, 53(13), 2892–2898.

GOI: Government of India 2018. Press information Bureau, Ministry of Agriculture & Farmers Welfare. http://pib.nic.in/newsite/PrintRelease.aspx?relid=175158.

Hashim, N., Janius, R.B., Baranyai, L., Rahman, R.A., Osman, A. and et al. (2012). Kinetic model for colour changes in bananas during the appearance of chilling injury symptoms. Food Bio Technol 5(8): 2952–2963.

Hashim, N., Pflanz, M., Regen, C., Janius, R. B., Abdul Rahman, R. and et al. (2013). An approach for monitoring the chilling injury appearance in bananas by means of backscattering imaging. J. Food Eng 116(1): 28–36.

Hu, D., Fu, X.P., Wang, A.C. and Ying, Y.B. 2015. Measurement methods for optical absorption and scattering properties of fruits and vegetables. Trans of the ASABE, 58(5):1387–1401

Jacques, S.L. (2013). Optical properties of biological tissues: A review. Physics in Medicine and Biology 58(11): 37–61.

Loeb, G. and Barton, J.K. (2003). Imaging botanical subjects with optical coherence tomography: A feasibility study. Trans. ASABE, 46(6), 1751-1758.

Lorente, D., Zude, M., Idler, C., Gómez-Sanchis, J. and Blasco, J. (2015). Laser-light backscattering imaging for early decay detection in citrus fruit using both a statistical and a physical model. J. Food Eng, 154: 76–85.

Lorente, D., Zude, M., Regen, C., Palou, L., Gómez- Sanchis, J. and Blasco, J. (2013). Early decay detection in citrus fruit using laser-light backscattering imaging. Postharvest Biol Technol 86: 424–430.

Lorenzo, J.R. (2012). Principles of Diffuse Light Propagation:Light Propagation in Tissues with Applications in Biology and Medicine. Hackensack, N.J.: World Scientific.

Lu, R. (2004b). Prediction of apple fruit firmness by near-infrared multispectral imaging. J. Texture Studies, 35(3): 263-276.

Lu, R., Cen, H., Huang, M. and Ariana, D. P. (2010). Spectral absorption and scattering properties of normal and bruised apple tissue. Trans. ASABE, 53(1): 263–269.

Mireei, S.A. (2010). Nondestructive determination of effective parameters on maturity of Mozafati and Shahani date fruits by NIR spectroscopy technique. PhD diss. Persian, Iran: University of Tehran, Department of Mechanical Engineering of Agricultural Machinery.

Neeraj, Chittora, A., Bisht, V. and Johar, V. (2017). Marketing and Production of Fruits and Vegetables in India. Int. J. Curr. Microbiol. App. Sci., 6(9): 2896–2907.

Nicolaï, B.M., Verlinden, B. E., Desmet, M., Saevels, S., Saeys, W., Theron, K., Cubeddu, R., Pifferi, A. and Torricelli, A. (2008). Time-resolved and continuous-wave NIR reflectance spectroscopy to predict soluble solids content and firmness of pear. Postharvest Biol. Tech., 47(1): 68–74.

Pfefer, T.J., Bennett, C.L., Durkin, A.J., Ediger, M.N., Gall, J.A., Matchette, L.S. and Wilke, J.N. (2003). Reflectance-based determination of optical properties in highly attenuating tissue. J. Biomed. Optics, 8(2): 206–215.

Prahl, S.A., van Gemert M.J.C., and Welch, A.J. (1993) Determining the optical properties of turbid media by using the adding-doubling method. Applied Optics, 32(4): 559–568.

Qin, J. and Lu, R. (2007). Measurement of the absorption and scattering properties of turbid liquid foods using hyperspectral imaging. Appl. Spectros., 61(4): 388-396.

Qin, J. and Lu, R. (2008). Measurement of the optical properties of fruits and vegetables using spatially resolved hyperspectral diffuse reflectance imaging technique. Postharvest Biol. Tech., 49(3): 355–365.

Rizzolo, A., Vanoli, M., Spinelli, L. and Torricelli, A. (2010). Sensory characteristics, quality, and optical properties measured by time-resolved reflectance spectroscopy in stored apples. Postharvest Biol. Technol., 58(1): 1–12.

Seifert, B., Zude, M., Spinelli, L., & Toricelli, A. (2015). Optical properties of developing pip and stone fruit reveal underlying structural changes. Physiologia Plantarum, 153(2), 327-336. http://dx.doi.org/10.1111/ppl.12232.

Tijskens, L., Zerbini, P.E., Schouten, R., Vanoli, M., Jacob, S., Grassi, M., Cubeddu, R., Spinelli, L. and Torricelli, A. (2007). Assessing harvest maturity in nectarines. Postharvest Biol. Tech., 45(2): 204–213.

Trong, N.N.D., Rizzolo, A., Herremans, E., Vanoli, M., Cortellino, G. and et al. (2014). Optical properties-microstructure-texture relationships of dried apple slices: Spatially resolved diffuse reflectance spectroscopy as a novel technique for analysis and process control. Innovative Food Sci. Emerging Tech., 21: 160–168.

Tuchin, V.V. (2000). TissueOptics: light scattering methods and instruments for medical diagnosis. SPIE Press, Bellingham,WA, USA.

Tuchin, V.V. (2000). Tissue optics: Light scattering methods and instruments for medical diagnosis. SPIE Tutorials texts in optical engineering TT 38 SPIE Press. Bellingham WA:

Tuchin, V.V. (2007). Tissue optics: Light scattering methods and instruments for medical diagnosis. 2nd edn. Vol. PM 166. Bellingham WA: SPIE Press.

Vanoli, M., Rizzolo, A., Grassi, M., Spinelli, L., Verlinden, B. E. and Torricelli, A. (2014). Studies on classification models to discriminate 'Braeburn' apples affected by internal browning using the optical properties measured by time-resolved reflectance spectroscopy. Postharvest Biol. Tech., 91: 112–121.

Verboven, P., Nemeth, A., Abera, M.K., Bongaers, E., Daelemans, D. and et al. (2013). Optical coherence tomography visualizes microstructure of apple peel. Postharvest Biol. Tech., 78: 123–132.

Wang, J., Liu, J. and Liu, T. (2015). The difference in effective light penetration may explain the superiority in photosynthetic efficiency of attached cultivation over the conventional open pond for microalgae. Biotechnol for Biofuels, 8:49.

Wang, W., and Li, C. (2013). Measurement of the light absorption and scattering properties of onion skin and flesh at 633 nm. Postharvest Biol. Tech., 86: 494–501.

Wang, W., Li, C. and Gitaitis, R.D. (2014). Optical properties of healthy and diseased onion tissues in the visible and near-infrared spectral region. Trans. ASABE, 57(6): 1771–1782.

Wilson, B.C., Patterson, M.S. and Flock, S.T. (1987). Indirect versus direct techniques for the measurement of the optical properties of tissues. Photochem. Photobiol., 46(5): 601–608.

Xu, H.R., Ying, Y.B., Fu, X.P. and Zhu, S.P. (2007). Near-infrared spectroscopy in detecting leaf miner damage on tomato leaf. Biosystems Engineering, 96(4): 447–454.

Zerbini, P.E., Grassi, M., Cubeddu, R., Pifferi, A. and Torricelli, A. (2002). Nondestructive detection of brown heart in pears by time-resolved reflectance spectroscopy. Postharvest Biol. Tech., 25(1), 87–97.

Zhang, L., Shi, A. and Lu, H. (2012). Determination of optical coefficients of biological tissue from a single integrating-sphere. J. Modern Optics, 59(2): 121–125,

Zhu, Q., He, C., Lu, R., Mendoza, F. and Cen, H. (2015). Ripeness evaluation of 'Sun Bright' tomato using optical absorption and scattering properties. Postharvest Biol. Tech., 103: 27–34.

Zidouk, A. and Styles, I. B. (2011). Using MAP to recover the optical properties of a biological tissue from reflectance measurements. Proc. SPIE, 8087: 80872A.

Zijp, R. and ten Bosch, J. J. (1998) Optical properties of bovine muscle tissue *in vitro;* a comparison of methods. Phys. Med. Biol., 43: 3065–3081.

10

Properties of Sem (*Dolichos lablab*)

Abhinay Shashank and Durga Shankar Bunkar

Introduction

Sem (*Dolichos*) bean, Hyacinth bean or Field bean is one of the most ancient crops among cultivated plants. It is a bushy, semi-erect, perennial herb, showing no tendency to climb. It is mainly cultivated either as a pure crop or mixed with finger millet, groundnut, castor, corn, bajra or sorghum in Asia and Africa (Maass *et al.*, 2010). It is a multipurpose crop grown for pulse, vegetable and forage. The crop is grown for its green pods, while dry as well as fresh green matured seeds are used in various vegetable food preparations, as most of the nutritive value is concentrated in the seeds. FAO website reports that *Dolichos lablab* (field bean) is a dual purpose legume. It is traditionally grown as a pulse crop for human consumption in south and south-east Asia and eastern Africa. Flowers and immature pods also used as a vegetable. It is also used as a fodder legume sown for grazing and conservation in broad acre agricultural systems in tropical environments with a summer rainfall. Also used as green manure, cover crop and in cut and carry systems and as a concentrate feed.

Gowda and Paul (2002) reported that this crop is also grown in home gardens as annual crop or on fences as perennial crop. It is one of the major sources of protein in the diets in southern states of India. The consumer preference varies with pod size, shape, colour and aroma (pod fragrance). It is also grown as an ornamental plant, mostly in USA for its beautiful dark-green, purple-veined foliage with large spikes clustered with deep-violet and white pea-like blossoms. Within India, Lablab is a field crop mostly confined to the peninsular region and cultivated to a large extent in Karnataka and adjoining districts of Tamil Nadu, Andhra Pradesh and Maharashtra. Karnataka contributes a major share, accounting for nearly 90 per cent in terms of both area and production in the country. Karnataka state records production of about 18,000 tonnes from an area of 85,000 hectares. Outside India, the crop is cultivated in East Africa, with similar uses, and in Australia as a fodder crop. Thus, Lablab bean is an important legume crop not only in India but also in other parts of the world.

Physical properties

Dolichos lablab seeds are enclosed in a pod like peas and bean crops. The physical properties of the seeds include various parameters like seed thickness, seed width, seed length, seed weight, seed color etc (Table 1). Subagio (2006) studied various parameters and reported them as follows:

Table 1 Physical properties of Dolichos lablab seeds (Subagio, 2006)

Physical properties	Value
Seed thickness (cm)	0.40 ± 0.03
Seed width (cm)	0.74 ± 0.05
Seed length (cm)	1.05 ± 0.10
10 Seeds volume (ml)	1.91 ± 0.399
Seed weight (g)	0.2334 ± 0.0287
Seed longitudinal surface area (cm^2)	0.8466 ± 0.1253
Seed husk thickness (mm)	0.10 ± 0.01
Percentage after de-coating (%)	83.2128 ± 1.1077
Seed colour	L = 73.4 ± 3.4; a* = 3.9 ± 1.4; b* = 9.3 ± 2.3; c* = 10.1 ± 1.2; H = 67.3° ± 4.6°

Nutritional properties

The nutritional composition of the *Dolichos lablab* seeds have been studied by various authors. Messina (1999) observed that throughout the world, an important role is played by legumes in the traditional diets of many regions. However, contrary to this, beans tend to play only a minor dietary role in western countries despite being low in fat and rich in protein, dietary fiber, and a variety of micronutrients and phytochemicals. Subagio (2006) reported the nutritional composition of the seeds on the wet basis: Moisture (9.3 ± 0.5 %), Protein (17.1 ± 1.5%), Lipid (1.1 ± 0.4%), Ash (3.6 ± 0.1%) and Carbohydrate (67.9 ± 4.2 %). Similarly, Kalpanadevi and Mohan (2013) studied two samples of seed materials of the underutilized tribal pulse, *D. lablab* var. *vulgaris* (dark brown and pale brown coloured seed coat) and reported the values in dry basis (seed powder). The study revealed that it has higher amounts of crude lipid and crude protein than most of the commonly consumed pulses and is also rich in minerals such as Na, K, P, Ca, Mg and Fe. The major bulk of seed protein is constituted by globulin and albumin.The seed lipids contain unsaturated fatty acids in the range of 66.78-69.08% and particularly linoleic acid in the range of

40.36-41.62%. The average of the proximate analysis of *Dolichos lablab* seeds have been tabulated in Table 2.

Table 2 The proximate analysis of Dolichos lablab seeds (Kalpanadevi and Mohan, 2013)

Components	% Dry weight basis
Moisture	7.39
Protein	21.82
Lipid	5.48
Total Dietary Fiber	6.45
Ash	4.14
Nitrogen Free Extractives	62.08
Calorific Value(KJ $100g^{-1}$ DM)	1607.65

Habib *et al.,* (2017) reported that the underutilized Kenyan variety of *Dolichos lablab* bean seeds serves as a good source of nutrients and contain many health-promoting components, such as fiber, proteins, minerals, and numerous phytochemicals endowed with useful biological activities, that allow it to contribute in a relevant way to the daily intake of these nutrients. Chau *et al.,* (1997) evaluated the effects of domestic cooking on the content of amino acids of *Dolichos lablab* seeds. Cooking of *Dolichos lablab* seeds for 60 min also resulted in a significant reduction ($P < 0{\cdot}05$) in the apparent recovery of all the essential amino acids except leucine, histidine, lysine and threonine. Increased cooking time *Dolichos lablab* (120 min) led to a lower apparent recovery of methionine (28·9–31·6%) and cystine (17·1–19·3%).

Antinutritional compounds

Dolichos lablab seeds contain a number of antinutritional factors which may beproteolytic inhibitors, phytohemagglutinins, lathyrogens, cyanogenetic compounds, compounds causing favism, factors affecting digestibility and saponins (Gupta, 1987). They includephenolics, tannins, L-DOPA, phytic acid, hydrogen cyanide, trypsin inhibitor, oligosaccharides (raffinose, stachyose, verbascose) and phytohaemagglutinating (Kalpanadevi and Mohan, 2013). The average of the values of *Dolichos lablab* seeds have been tabulated in Table 3.

Subagio, *et al.,* (2006) reported that, since the contents of phytate and trypsin inhibitor are 18.9 ± 0.2 mg/g and 0.15 ± 0.02 TIU/mg, respectively, some treatments are needed to reduce their antinutritional factors, before using the seeds as food. Some common treatments are soaking, germination, boiling, pressure cooking,cooking and autoclaving. Ramakrishna *et al.* (2008) did the comparative analysis of changes in anti-nutritional factors by germination,

Table 3 Anti-nutritional factors of Dolichos lablab seeds (Kalpanadevi and Mohan, 2013)

Anti-nutritional factors	% Dry weight basis
Total free phenolics g /100 g	0.57
Tannins g /100g	0.19
L-DOPAg /100g	0.87
Phytic acid g /100g	388.21
Hydrogen cyanide mg /100g	0.24
Trypsin inhibitor (TIA mg^{-1} protein)	31.24

boiling, roasting and pressure cooking of the *Dolichos lablab* L. seeds. The raw dry Indian bean had a very high TIA which decreased progressively to 51% after 12 h soaking and upto 17% at 32 h germination period. An overall fall in polyphenols, tannins, phytic acids and phytate phosphorus of 70%, 46%, 36% and 30% was noticed during the 32h germination period. Roasting caused a maximum reduction was observed in TIA and phytic acids with, while the boiling and pressure cooking decreases the levels of polyphenols and tannins. The study concludes that germination is a more effective method in reducing TIA, tannins, polyphenols and phytic acid than the various cooking treatments. Similar to this, Osman *et al.,* (2007) studied the changes in nutrient composition of trypsin inhibitors, phytic acid, tannins, and in-vitro digestibility of protein of lablab bean during 5 days of germination. The crude protein was significantly increased, whereas lipid and carbohydrates content decreased. Antinutritional study revealed that trypsin activity and phytic acid content decreased as germination period increased. In contrast, there was progressive increase in germination time. *In-vitro* protein digestibility markedly increased with germination time, with a significant increase in day 5.

Similar to this, Chau *et al.,* (1997) observed that cooking for 60 min was found to be effective in reducing the tannin contents of *Dolichos lablab* seeds by 74·6%, respectively. Upon cooking, phytate and trypsin inhibitory activity in different seeds were also reduced to different extents depending on the cooking times. Vijayakumari *et al.,* (1995) investigated the effects of soaking, cooking and autoclaving on changes in polyphenols, phytohaemagglutinating activity, phytic acid, hydrogen cyanide (HCN), oligosaccharides and in vitro protein digestibility in seeds of *Dolichos lablab* var. *vulgaris*. Both distilled water and $NaHCO_3$ solution soaking and autoclaving significantly reduced the contents of total free phenolics (85–88%) compared to raw seeds. Autoclaving (45 min) reduced the content of tannins by upto 72%. Soaking seemed to have limited effect in eliminating phytohaemagglutinating activity, whereas autoclaving (45 min) seemed to eliminate the haemagglutinating activity completely. The reduction in content of phytic acid was found to be somewhat greater in distilled water soaking (28%) compared to $NaHCO_3$ solution soaking (22%). Only a limited loss in content of phytic acid was observed under cooking as well as autoclaving. Loss of HCN was greater under autoclaving (87%) compared to the other processes studied. Of the three sugars analysed, soaking reduced the level of verbascose more than that of stachyose and raffinose. Autoclaving reduced the content of oligosaccharides more efficiently (67–86%) than ordinary cooking (53–76%). Autoclaving improved the in vitro protein digestibility (IVPD) significantly (13%). Of all the different water and hydrothermal treatments studied autoclaving seemed to be the most efficient method in improving IVPD and eliminating the antinutrients investigated except phytic acid. Contrary to these, Maheshu *et al.*, (2013) reported that the dry heating caused remarkable increase in tannin contents (1.809±0.25 g GAE/ 100 g extract)

Bioactive and functional compounds in Dolichos beans

Dolichos bean seeds contain a number of bioactive and functional compounds like phenols, sterols, vitamins, pigments, lectins, brassinolide, dolicholide, castasterone, Dolichosterone and homodolichosterone etc. (Baba *et al.*, 1983; Yokota *et al.*, 1983). Al-Snafi (2017) also reported the presence of sugar, alcohols, phenols, steroids, essential oils, alkaloids, tannins, flavonoids, saponins, coumarins, terpenoids, pigments, glycosides, anthnanoids, wide range of minerals and many other metabolites in *Dolichos lablab*seeds. Lectins are the diverse class of carbohydrate interacting proteins, having great potential as immunopotentiating and anti-cancer agents (Sitienei *et al.*, 2017).

Mo *et al.*, (1999) purified the mannose/glucose-binding *Dolichos lablab* lectin (designated DLL) from seeds of *Dolichos lablab.* The purified lectin gave a single symmetric protein peak with an apparent molecular mass of 67 kDa on gel filtration chromatography, and five bands ranging from 10 kDa to 22 kDa upon SDS–PAGE. He concluded that the *Dolichos lablab* lectin has neither an extended carbohydrate combining site, nor a hydrophobic binding site adjacent to it. DLL strongly precipitates murine IgM but not IgG, and the recent finding that this lectin interacts specifically with NIH 3T3 fibroblasts transfected with the Flt3 tyrosine kinase receptor and preserves human cord blood stem cells and progenitors in a quiescent state for prolonged periods in culture, make this lectin a valuable tool in biomedical research.

Paul *et al.*, (2000) purified the polyphenol oxidase from field bean (*Dolichos lablab*) seeds to apparent homogeneity by a combination of ammonium sulfate precipitation, DEAE-Sephacel chromatography, phenyl agarose chromatography, and Sephadex G-200 gel filtration. The purified enzyme has a molecular weight of 120±3 kDa and is a tetramer of 30±1.5 kDa. Native polyacrylamide gel electrophoresis of the purified enzyme revealed the presence of a single isoform with an observed pH optimum of 4.0. 4-Methyl catechol is the best substrate, followed by catechol, and L-3, 4-dihydroxyphenylalanine, all of which exhibited a phenomenon of inhibition by excess substrate. No activity was detected toward chlorogenic acid, catechin, caffeic acid, gallic acid, and monophenols. Tropolone, both a substrate analogue and metal chelator, proved to be the most effective competitive inhibitor with an apparent K_i of 5.8×10^{-7} M. Ascorbic acid, metabisulfite, and cysteine were also competitive inhibitors.

Gowda and Paul (2002) studied the hydroxylation of ferulic acid and tyrosine by field bean (*Dolichos lablab*) polyphenol oxidase, a reaction that does not take place without the addition of catechol. A lag period similar to the characteristic lag of tyrosinase activity was observed, the length of which decreased with increasing catechol concentration and increased with increasing ferulic acid concentration. The presence of o-diphenols, viz. catechol, L-dihydroxyphenylalanine, and 4-methyl catechol, is also necessary for the oxidation of the diphenols, caffeic acid, and catechin to their quinones by the field bean polyphenol oxidase. This oxidation reaction occurs immediately with no lag period and does not occur without the addition of diphenol. The absence of a lag period for this reaction indicates that the diphenol mechanism of diphenolase activation differs from the way in which the same o-diphenols activate the monophenolase activity.

Health Benefits

Dolichos lablab seeds contain a number of nutritional, bioactive and functional compounds because of which it exhibits various health benefits. Al-Snafi (2017) reviewed various preliminary pharmacological studies which showed that *Dolichos lablab* possessed antidiabetic, antiinflammatory, analgesic, antioxidant, cytotoxic, hypolipidemic, antimicrobial, insecticidal, hepatoprotective, antilithiatic, antispasmodic effects and also used forthe treatment of iron deficiency anemia.

Antioxidative: Marathe *et al.,* (2011) did a comparative study on antioxidant activity of different varieties of commonly consumed legumes in India using radical scavenging [(2,20-azino-bis (3-ethylbenz-thiazoline-6-sulfonic acid, (ABTS) and (1,1-diphenyl-2-picryl-hydrazyl (DPPH)], metal ion (Fe^{2+}) chelation assays and Ferric Reducing Antioxidant Power (FRAP). Horse gram, common beans, cowpea (brown and red) and fenugreek showed high DPPH radical scavenging activity (>400 units/g), while lablab bean (cream and white), chickpea (cream and green), butter bean and pea (white and green) showed low antioxidant activity (<125 units/g). Green gram, black gram, pigeon pea, lentils, cowpea (white) and common bean (maroon) showed intermediate activity. Similar trend was observed when the activity was assessed with ABTS and FRAP assays. Antioxidant activity showed positive correlation ($r^2 > 0.95$) with phenolic contents, in DPPH, ABTS and FRAP assays, whereas poor correlation ($r^2 = 0.297$) was observed between Fe^{2+} chelating activity of the legumes and phenolic contents.

Maheshu *et al.,* (2013) studied the effects of raw, dry heated and pressure cooked samples on total phenolic components and antioxidant activity in commonly consumed field bean, *Dolichos lablab* L. was investigated. The raw and processed samples were extracted with 70% methanol. Dry heated samples of *D. lablab* was found to possess the highest DPPH (IC50, 2.53±0.17 μg/ml), TEAC (4649.8±38.4 μmol/g DM), $OH^{\cdot}$ radical (IC50, 42.2±0.67 μg/ml) scavenging activities, inhibition of linoleic acid and ferric reducing capacity than other samples. The raw samples displayed the highest antihemolytic activity (59.6±1.53%) and chelating capacity (74.2±1.37 mg EDTA/g). Dry heat processing exhibited several advantages in retaining the antioxidant components and activities. The higher correlation was found the phenolic content with chelating (r^2=0.933) and antihemolytic (r^2=0.839) activities, but a poor correlation with other assays. Moreover, the content of tannins gave good correlation (r^2=0.644–0.997) with all antioxidant assays. The low correlation values between total phenols and the antioxidative activity suggest that the major antioxidant compounds in studied seeds might be tannins.

Antihyperglycemic Property: Ahmed *et al.,* (2015) found out that administration of methanol extract of beans led to dose-dependent and significant reductions in blood glucose levels in glucose-loaded mice. At doses of 50, 100, 200 and 400 mg per kg body weight, the extract reduced blood glucose levels by 16.4, 39.1, 40.1, and 54.8%, respectively compared to control animals. They concluded that the fruits can be used as a source for lowering blood sugar in diabetic patients.

Antinociceptive Property: Ahmed *et al.,* (2015) studied antinociceptive properties of methanol extract of *Dolichos lablab* beans. In antinociceptive activity tests, the extract at the above four doses reduced the number of abdominal constrictions by 32.3, 45.2, 54.8, and 58.1, respectively. They concluded that the fruits can be used as a source for alleviating pain.

Anticancerous Property: Sitienei *et al.,* (2017) reported that Lectins are the diverse class of carbohydrate interacting proteins, having great potential as immunopotentiating and anti-cancer agents. The present investigation sought to demonstrate the anti-proliferative activity of *Dolichos lablab* lectin (DLL) encompassing immunomodulatory attributes. DLL specific to glucose and mannose carbohydrate moieties has been purified to homogeneity from the common dietary legume *D. lablab.* Results elucidated that DLL agglutinated blood cells non-specifically and displayed striking mitogenicity to human and murine lymphocytes *in vitro* with interleukin (IL)-2 production. The DLL-conditioned medium exerted cytotoxicity towards malignant cells and neoangiogenesis*in vitro.* Similarly, *In vivo* anti-tumour investigation of DLL elucidated the regressed proliferation of ascitic and solid tumour cells, which was paralleled with blockade of tumourneovasculature. DLL-treated mice showed an up-regulated immunoregulatory cytokine IL-2 in contrast to severely declined levels in control mice. Mechanistic validation revealed that DLL has abrogated the microvessel formation by weakening the proangiogenic signals, specifically nuclear factor kappa B (NF-*κB), hypoxia inducible factor 1α (HIF-*1 *α*), matrix metalloproteinase (MMP)-2 and 9 and vascular endothelial growth factor (VEGF) in malignant cells leading to tumour regression. In summary, it is evident that the dietary lectin DLL potentially dampens the malignant establishment by mitigating neoangiogenesis and immune shutdown. For the first time, to our knowledge, this study illustrates the critical role of DLL as an immunostimulatory and anti-angiogenic molecule in cancer therapeutics.

Value Added Products

Dolichos lablab seeds being one of the most nutritious bean crop shows potential to be used as various value added products apart from its traditional uses as vegetable and animal feed.

Dolichos lablab flour: Chau and Cheung (1998) investigated the functional properties of *Phaseolus angularis, Phaseolus calcaratus* and *Dolichos lablab* flours and compared with those of soybean flour (Table 4). Cleaned seeds weremanually dehulled after soaking in distilled water atroom temperature (24°C) (10 h for *P. angularis* and *P. calcaratus* seeds and 3 h for *D. lablab* seed and soybean).The cotyledons were freeze-dried and then ground in acyclotec mill to pass through a 0.5 mm screen and defatted with acetone.The minimum nitrogen solubilities of *D. lablab* flour was at pH 4 compared to pH 5 for that of *P. angularis* and *P. calcaratus* flours. Compared with soybean flour, *P. angularis, P calcaratus* and *D. lablab* flours exhibited lower foam capacities, water- and oil-holding capacities, but higher gelation capacities. The emulsifying activities and emulsion stabilities of all legume flours tested were pH-dependent with minimum values at pH 4. Their emulsion stabilities were greater than 80.2% from pH 2 to 10, except at pH 4. Foam capacities and stabilities were also pH-dependent, highest foam stabilities being at pH 4.

Table 4 Comparison of functional properties of some flours (Chau and Cheung, 1998)

Flours	Bulk density ($g\ ml^{-1}$)	pH Value	Least Gelation concentration (%)	Water Holding Capacity (%)	Oil Holding Capacity (%)
P. angularis	0.62	6.60	12.0	1.46	1.40
P. calcaratus	0.59	6.53	13.0	1.47	1.28
D. lablab	0.64	6.59	10.0	1.04	1.19
Soybean	0.47	6.65	17.0	1.80	1.93

Protein isolate: Subagio (2006) prepared protein isolate from the seeds using an isoelectric method, which was also used to characterize the physicochemical and functional properties. Hyacinth bean seeds have a moderate concentration of protein (17.1 ± 1.5%), and low concentration of HCN (1.1 ± 0.1 mg/100 g). Using the isoelectric preparation, the yield of protein isolate was low (7.38 ± 0.2 g per 100 g of the seeds), but the protein isolate had good colour, neutral odour, high protein content (89.8 ± 0.82%), and low ash (2.97 ± 0.36%). The protein isolate also had good functional properties, such as foaming capacity (232 ± 12.2ml/g), solubility, and emulsifying activity index (534 ± 4.5 m^2/g). However, the foaming stability (2.3 ± 0.2 min) and emulsifying stability index (2.7 ± 0.1 h) were low.

Shelf-stable green legumes: Bhattacharya and Malleshi (2012) reported that premature green legumes are good sources of nutraceuticals and antioxidants and are consumed as snacks as well as vegetables. They processed Bengal gram (*Cicer arietinum*) and field bean (*Dolichos lablab*) to prepare shelf-stable green legumes (SSGL) to extend their availability throughout the year. The shelf stable green legumes (SSGL) show higher water absorption capacity compared to matured dry legumes (MDL). The total colour change in the processed/dried SSGL and MDL samples increased significantly ($p \leq 0.05$) compared to the freshly harvested green samples. The carotenoid content of Bengal gram and field bean SSGLs are 8.0 and 3.2 mg/100 g, and chlorophyll contents are 12.5 and 0.5 mg/100 g, respectively, which are in negligible quantities in matured legumes; the corresponding polyphenol contents are 197.8 and 153.1 mg/100 g. These results indicate that SSGLs possess potential antioxidant activity.

References

Ahmed, M., Trisha, U.K., Shaha, S.R., Dey, A.K. and Rahmatullah, M. (2015). An initial report on the antihyperglycemic and antinociceptive potential of Lablab purpureus beans. World Journal of Pharmacy and Pharmaceutical Sciences, 4(10): 95–105.

Al-Snafi, A.E. (2017). The pharmacology and medical importance of Dolichos lablab (*Lablab purpureus*) - A review. IOSR Journal of Pharmacy, 7(2): 22–30.

Baba, J., Yokota, T. and Takahashi, N. (1983). Brassinolide-related new bioactive steroids from Dolichos lablab seed. Agricultural and Biological Chemistry, 47(3): 659–661.

Bhattacharya, S. and Malleshi, N.G. (2012). Physical, chemical and nutritional characteristics of premature-processed and matured green legumes. Journal of Food Science and Technology, 49(4): 459–466.

Chau, C.F. and Cheung, P.C.K. (1998). Functional properties of flours prepared from three Chinese indigenous legume seeds. Food Chemistry, 61(4): 429–433.

Chau, C.F., Cheung, P.C.K. and Wong, Y.S. (1997). Effects of cooking on content of amino acids and antinutrients in three Chinese indigenous legume seeds. Journal of the Science of Food and Agriculture, 75(4): 447–452.

Gowda, L.R. and Paul, B. (2002). Diphenol activation of the monophenolaseanddiphenolase activities of field bean (Dolichos lablab) polyphenol oxidase. Journal of Agricultural and Food Chemistry, 50(6): 1608–1614.

Gupta, Y.P. (1987). Anti-nutritional and toxic factors in food legumes: a review. Plant Foods for Human Nutrition, 37(3): 201–228.

Habib, H.M., Theuri, S.W., Kheadr, E.E. and Mohamed, F.E. (2017). Functional, bioactive, biochemical, and physicochemical properties of the Dolichos lablab bean. Food & Function, 8(2): 872–880.

Kalpanadevi, V. and Mohan, V.R. (2013). Nutritional and anti nutritional assessment of under utilized legume Dolichos lablab var. vulgaris. Bangladesh Journal of Scientific and Industrial Research, 48(2): 119–130.

Maass, B.L., Knox, M.R., Venkatesha, S.C., Angessa, T.T., Ramme, S. and Pengelly, B.C. (2010). Lablab purpureus — a crop lost for Africa? Tropical plant biology, 3(3): 123–135.

Maheshu, V., Priyadarsini, D.T. and Sasikumar, J.M. (2013). Effects of processing conditions on the stability of polyphenolic contents and antioxidant capacity of Dolichos lablab L. Journal of Food Science and Technology, 50(4): 731–738.

Marathe, S.A., Rajalakshmi, V., Jamdar, S.N. and Sharma, A. (2011). Comparative study on antioxidant activity of different varieties of commonly consumed legumes in India. Food and Chemical Toxicology, 49(9): 2005–2012.

Messina, M.J. (1999). Legumes and soybeans: overview of their nutritional profiles and health effects. The American Journal of Clinical Nutrition, 70(3): 439–450.

Mo, H., Meah, Y., Moore, J.G. and Goldstein, I.J. (1999). Purification and modeling of gas exchange in modified atmosphere packaging. Journal of Food and Process Engineering, 34(1): 239–245.

Osman, M.A. (2007). Changes in nutrient composition, trypsin inhibitor, phytate, tannins and protein digestibility of Dolichos lablab seeds (Lablab purpuresus (L) sweet) occurring during germination. Journal of Food Technology, 5(4): 294–299.

Ramakrishna, V., Rani, P.J. and Rao, P.R. (2008). Changes in anti-nutritional factors in Indian bean (*Dolichos lablab* L.) seeds during germination and their behaviour

during cooking. Nutrition and Food Science, 38(1): 6–14. DOI: http://dx.doi.org/10.1108/00346650810847963
Sitienei, R.C., Onwonga, R.N., Lelei, J.J. and Kamoni, P. (2017). Use of Dolichos (Lablab Purpureus L.) and combined fertilizers enhance soil nutrient availability, and maize (Zea Mays L.) yield in farming systems of Kabete Sub County, Kenya. Agric. Sci. Res. J., 7: 41–61.
Subagio, A. (2006). Characterization of hyacinth bean (*Lablab purpureus* (L.) sweet) seeds from Indonesia and their protein isolate. Food chemistry, 95(1): 65–70.
Vijayakumari, K., Siddhuraju, P. and Janardhanan, K. (1995). Effects of various water or hydrothermal treatments on certain antinutritional compounds in the seeds of the tribal pulse, Dolichos lablab var. vulgaris L. Plant Foods for Human Nutrition (Formerly Qualitas Pantarum), 48(1): 17–29.
Yokota, T., Baba, J. and Takahashi, N. (1983). Brassinolide-related bioactive sterols in Dolichos lablab: brassinolide, castasterone and a new analog, homodolicholide. Agricultural and Biological Chemistry, 47(6): 1409–1411.

11

Physical Properties of Food Materials

Ashish M. Mohite and Neha Sharma

Abstract

The chapter aims at introducing selected physical properties of food and biological materials in food process and application in the field of food technology. These physical properties of food are not only used for designing of food equipments but can also maintain the nutritive value of food to a greater extent during processing in these equipments.

Introduction

The investigation of food engineering is spotlighted on the inspection of equipments and system used to process food materials on a commercial scale. Building of system for food materials can be efficient if two factors are known, the nature of the food product and what happen to the food during its processing. Food materials are organic in nature and it has a certain kind of qualities which recognize them from other processed food product. Since food materials are organic in nature, they have (a) shapes usually found in normally non regular; (b) properties which differ from other food products; (c) heterogeneous arrangement; (d) properties are varied due to climatic conditions, ripening conditions, etc. and (e) several other factors such as (e) influenced by chemical changes, moisture content variation and enzymatic movement. Managing food materials with these remarkable attributes requires extra thought, for the most part by implication and extra sources or reasons for different varieties. The attributes of a food material that are autonomous of the on looker, quantifiable, can be measured, and characterize the condition of the material (yet not how it accomplished that state) are considered as its physical properties.

Physical properties are the one of a kind of characteristic way of food material reacts to physical behavior, including mechanical, thermal, electrical, optical, and electromagnetic properties. A superior comprehension of the manner in which food materials react to physical and chemical properties takes into account the ideal planning, food equipment and procedures to guarantee food quality and safety. Usually for the physical properties of a food material the change occurs due to handling activities. People unfamiliar with this natural variability of biological materials may overlook these factors or be frustrated by lack of control over the input parameters. The characteristics of a food material that are independent of the observations, measurable, can be quantified, and define the state of the material which is needed during processing. Knowledge of a food's physical property is necessary for: i) defining and quantifying a description of the food material, ii) providing basic data for food engineering and unit operations. It is common for the physical properties of a food to change during processing operations. Not recognizing these changes can lead to potential processing failures. The select physical properties of food are introduced in this chapter.

Shape

Mohsenin (1970) stated that roundness, by definition, is a measure of the sharpness of the corners of the solid. Roundness is under different conditions of geometry and application of particular foods. The shape description is as follows:

1. Round - Approaching spheroid
2. Oblate - Flattened at the stem end and peak
3. Oblong - Vertical breadth more noteworthy than the level distance across
4. Conic - Tapered toward the pinnacle
5. Ovate - Egg-molded and wide at the stem end
6. Oblate - Inverted oblate
7. Lopsided - Axis interfacing stem and summit inclined
8. Elliptical - Approaching ellipsoid
9. Truncate - Having the two closures squared or straightened
10. Unequal - One half bigger than the other parts.

Size

Specific gravity may be calculated from the size. Size is an important physical attribute of foods used in screening solids to separate foreign materials, For example, particle size of powdered milk must be large enough to prevent agglomeration, but small enough to allow rapid dissolution. It is easy to specify size for regular particles, but for irregular particles the term size must be arbitrarily specified. Sphericity expresses the characteristic shape of a solid object relative to that of a sphere. Interstitial air spaces have different values of particle density and bulk density. Sphericity communicates the trademark state of a strong item in respect to that of a circle. Materials witshout inner air spaces, for example, liquids and solids, have equivalent molecule and mass thickness. Molecule thickness is the mass isolated by the volume of the molecule alone. Mass thickness is the mass of a gathering of individual particles separated by the space involved with the whole mass, including the air space. Thickness of materials is helpful in scientific change of mass to volume. Particle sizes are expressed in different units depending on the size range involved. Coarse particles are measured in millimeters, fine particles in terms of screen size, and very fine particles in micrometers or nanometers.

Ultrafine particles are some of the time portrayed as far as their surface zone per unit mass, for the most part in square meters per gram (Coskuner and Karababa, 2007). Size can be resolved utilizing the anticipated territory technique. In this strategy, three trademark measurements are characterized:

1. Real breadth, which is the longest component of the most extreme anticipated region;
2. Halfway width, which is the base measurement of the most extreme anticipated zone or the greatest distance across of the base anticipated territory; and
3. The Minor distance across, which is the briefest component of the base anticipated region.

$$\text{Spherecity} = \frac{\text{Volume of solid sample}}{(\text{Volume of circumscribed sphere})^{1/3}}$$

Volume

The liquid volume is computed by determining the mass of the displaced water and dividing by the known density of the water. The mass of the displaced water is the scale's reading with the object submerged minus the mass of the container and water. For objects that float, it is necessary to force the object entirely into the water with a thin stiff rod. If the object is heavier than water, it must be suspended in the water by a rod or other support to insure that the added mass of the object. The following expression is used to calculate the volume of displaced water: Volume (m^3) = The specific gravity is defined as the ratio of the mass of that product to the mass of an equal volume of water at 4°C, the temperature at which water density is greatest.

Liquid Displacement Method: If the solid sample does not absorb liquid very fast, the liquid displacement method can be used to measure its volume. In this method, volume of food materials can be measured by pycnometers (specific gravity bottles) or graduated cylinders. The pycnometer has a small hole in the lid that allows liquid to escape as the lid is fitted into the neck of the bottle. The bottle is precisely weighed and filled with a liquid of known density.

The lid is placed on the bottle so that the liquid is forced out of the capillary. A liquid that has been forced out of the capillary is wiped from the bottle and the bottle is weighed again. After the bottle is emptied and dried, solid particles are placed in the bottle and the bottle is weighed again. The bottle is completely filled with liquid so that liquid is forced from the hole when the lid is replaced. The bottle is reweighed and the volume of solid particles can be determined from the following formula:

Vs = Weight of the liquid displaced by solid

Density of liquid = $(Wpl - Wp) - (Wpls - Wps) / \rho l$

where

Vs = volume of the solid (m^3),

Wpl = weight of the pycnometer filled with liquid (kg),

Wp = weight of the empty pycnometer (kg),

$Wpls$ = weight of the pycnometer containing the solid sample and filled with liquid (kg),

Wps = weight of the pycnometer containing solid sample with no liquid (kg),

$\rho 1$ = density of the liquid (kg/m^3).

The volume of a sample can be measured by direct measurement of the volume of the liquid displaced by using a graduated cylinder or burette. The difference between the initial volume of liquid in a graduated cylinder and the volume of liquid with immersed material gives us the volume of the material. That is, the increase in volume after addition of solid sample is equal to the solid volume. In the liquid displacement method, liquids used should have a low surface tension and should be absorbed very slowly by the particles. Most commonly used fluids are water, alcohol, toluene, and tetrachloroethylene.

Surface Area and Seed Dimensions

Perimeter: This is the length of the edge of a molecule's anticipated picture of the 2-D plane.

Maximum length: This is characterized as the most extreme length of a straight line associated with two on the edge of a molecule.

Maximum width: This is characterized as the most extreme length of a straight line opposite to the line of greatest length.

Unit volume: Unit volume of 100 individual food stuff (solids) can be calculated from the value of three axial dimensions (length, breath and thickness) (Murthy and Bhattacharya, 1998).

Surface area of food material can be found by analogy with a sphere of same geometric mean diameter (Sacilik et al., 2003). Specific surface area (surface area per unit weight or per unit volume) an important factor for processes of food materials it involves mass transfer through surfaces, e.g. respiration of fruits, extraction of coffee beans, smoking of hams, salting of cheeses, gas and water vapor transfer into and out of packages. Determining the ratio of usable material to peel in fruits and vegetables. This is important economically; one ton of small potatoes, for example, has a much greater total specific surface area (m^2/kg) than one ton of large potatoes, and therefore the wastage when they are peeled is much greater.

One thousand seed weight: The mass of seeds was determined on 100 randomly selected seeds and converted to a 1000-seed-basis.

Density

Density of food materials can also ensure the quality of food products. The unit operation such as separation process which includes centrifugation and sedimentation where pneumatic transport of powders and particulates of food products is required here densities has major importances. For liquid food materials transportation the density of liquid is required to determine the power required for pumping.

Solid density is the density of the solid material (including water), excluding any interior pores that are filled with air. It can be calculated by dividing the sample weight by solid volume determined by the gas displacement method in which gas is capable of penetrating all open pores up to the diameter of the gas molecule.

Material density is the density of a material measured when the material has been broken into pieces small enough to be sure that no closed pores remain.

Particle density is the density of a particle that has not been structurally modified. It includes the volume of all closed pores but not the externally connected ones. It can be calculated by dividing the sample weight by particle volume determined by a gas pycnometer.

Apparent density is the density of a substance, including all pores within the material (internal pores). Apparent density of regular geometries can be determined from the volume calculated using the characteristic dimensions and mass measured. Apparent density of irregularly shaped samples may be determined by solid or liquid displacement methods.

Bulk density is the density of a material when packed or stacked in bulk. Bulk density of particulate solids is measured by allowing the sample to pour into a container of known dimensions. It depends on the solid density, geometry, size, surface properties, and the method of measurement. The bulk density is the ratio of the mass sample of the seeds to its total volume. It was determined by filling a 1000 ml container with seeds from a height of about 15cm, striking the top level and weighing the contents (Mohite *et al.*, 2018).

True density is defined as the ratio of mass of the sample to its true volume, was determined using the toluene displacement method in order to avoid absorption of water during experiment. Fifty milliliter of toluene was placed in a 100 ml graduated measuring cylinder and 5 g seeds were immersed in the toluene (Barnwal *et al.*, 2015). The amount of displaced toluene is recorded from the graduated scale of the cylinder. The ratio of weight of seeds to the volume of displaced toluene gives the true density.

Porosity

Porosity is the percentage of air between the particles compared to a unit volume of particles. A frequently used method of measuring the volume of non-porous objects such as vegetables and fruits is the use of platform scales or a top loading balance to determine the volume of a displaced liquid such as water. Porosity is the level of air between the particles contrasted with a unit volume of particles. Porosity permits gases, for example, air, and fluids to move through a mass of particles alluded to as a pressed bed in drying and refining activities. Beds with low porosity (low rate air space) are increasingly impervious to liquid stream and in this manner are progressively hard to dry, warmth, or cool. With high porosity, wind, streams effectively through the bed, drying is quick, and the power required by fans and siphons is low. An as often as possible utilized strategy for estimating the volume of non-permeable articles, for example, vegetables and natural products is the utilization of stage scales or a top stacking equalization to decide the volume of a dislodged fluid, for example, water. The fluid volume is registered by deciding the mass of the dislodged water and separating by the known thickness of the water. The porosity is the fraction of the space in the bulk grain, which is not occupied by the grain. The porosity of the bulk seed was calculated from the values of true density and bulk density using the equation:

$$\varepsilon = \left(1 - \frac{\rho_b}{\rho_t}\right)$$

Where: ε is porosity, ρ_t is the true density and ρ_b is the bulk density.

Coefficient of static friction

The coefficient of static friction can be determine with respect to four surfaces: plywood, stainless steel, aluminum sheet, and galvanized iron. These are common materials used for transportation, storage and handling of grains, pulses and seeds construction of storage and drying bins. To determine the coefficient of static friction, a hollow metal cylinder 50 mm diameter and 50 mm high and open at both ends was used. The cylinder was filled with the seeds and placed on an adjustable tilting table. The surface was raised up gradually by a screw device until the cylinder just starts to slide down. The angle of the surface was read from a scale and the static coefficient of the friction was read as the tangent of this angle (Jain and Bal, 1997).

Angle of repose

Angle of repose can be determined using, a box measuring 300 mm × 300 mm × 300 mm, having a removable front panel. The box is filled with the seeds, and the front panel was quickly removed, allowing the seeds to flow to their natural slope. The angle of repose was calculated from measurements of seed free surface depths at the end of the box and midway along the sloped surface and horizontal distance from the end of the box to this midpoint (Mohite *et al.*, 2019; Ozturk and Esen, 2008).

Conclusion

Based on the above finding, the conclusions can be drawn that engineering properties differ from seeds to grain to fruits to vegetables. Shape is an important criteria and physical dimensions are important in screening solids to separate foreign materials from seeds and grains. The quality differences in grains and seeds can often be detected by differences in densities. Volumes and surface areas of seeds must be known for accurate modeling of heat and mass transfer during cooling and drying. The porosity affects the resistance to airflow through bulk solids. The static coefficient of friction and angle of repose is necessary to design conveying machine and hoppers which are used in planter machines.

References

Barnwal, P., Singh, K.K. Mohite, A.M., Sharma, A. and Saxena, S.N. (2015). Influence of cryogenic and ambient grinding on grinding characteristics of fenugreek powder: a comparative study. Journal of Food Processing and Preservation, 39(6): 1243–1250.

Coskuner, Y., and Karababa, E. (2007). Physical Properties of Coriander Seeds (*Coriandrum sativum* L.). Journal of Food Engineering, (80): 408–416.

Jain, R.K., and Bal, S. (1997). Properties of pearl millet. Journal of Agricultural Engineering Research, 66(2): 85–91.

Mohite, A.M., and Sharma, N. (2018). Drying behaviour and engineering properties of lima beans. Agricultural Engineering International: CIGR Journal, 20(3): 180–185

Mohite, A.M., Sharma, N. and Mishra, A. (2019). Influence of different moisture content on engineering properties of tamarind seeds. Agricultural Engineering International : CIGR Journal, 6(2): 220–224.

Mohsenin, N.N. (1970). Physical Properties of Plant and Animal Materials-II. London: Gordon and Breach Science Publishers.

Murthy, C.T., and Bhattacharya, S. (1998). Moisture dependent physical and uniaxial compression properties of black Pepper. Journal of Food Engineering, 37(2): 193–205.

Ozturk, T., and Esen, B. (2008). Physical and mechanical properties of barley. Agriculture Tropica ET Subtropica, 41(3): 117–121.

Sacilik, K., Özturk, R. and Keskin, R. (2003). Some physical properties of hemp seed. Biosystems Engineering, 86(2): 191–198.

12

Comparative Study on Physical Properties of Different Varietal Tomatoes

Tarun Kumar, Suresh Chandra, Samsher, Neelesh Chauhan
Jaivir Singh, Ankur M Arya, Kavindra Singh, Ratnesh Kumar

Abstract

During present investigation a comparative study was done to see the effect of storage on physical properties of tomatoes (Lycopersicon esculentum Mill) packed in HDPE and stored under ambient temperature. Four varieties of tomatoes taken for current investigation were Badshah, Himshikhar, NS-524 and Raja. From the studies it was clear that TSS increased gradually during storage duration for entire tomato varieties due to loss of some moisture from tomatoes. Variety NS-524 shows maximum shrinkage among all during observation period.

Introduction

Tomato (*Lycopersicon esculentum* Mill) is rich source of vitamins A, C, potassium, minerals and fibers. Tomatoes are rich in lycopene, a carotenoid which is important because of its health related properties. The effect of dietary lycopene in reducing the risk of chronic diseases, such as cancer and coronary heart diseases, has already been indicated in epidemiological studies (Giovannucci, 1999; Rao and Agarwal, 1999). It is one of the most important protective food crops of India. Use of tomatoes is increasing day by day and a variety of products like puree, syrup, paste, ketchup, juice etc. are made (Kumar *et al.*, 2018). It is grown in 0.458 M ha area with 7.277 M mt production and 15.9 mt/ha productivity. The major tomato producing states are Andhra Pradesh, Karnataka, Orissa, Maharashtra, Madhya Pradesh, West Bengal, Bihar and Uttar Pradesh (NCPAH, 2010). To handle such a large quantity of tomatoes; machine are needed and to design and optimization a machine for handling, cleaning, conveying, and storing, the physical attributes and their relationships must be known (Mirzaee *et al.*, 2008). As said by

Taheri-Garavand *et al.*, (2009) the physical properties of tomato are important to design the equipment for processing, transportation, sorting, separation and storing. Designing such equipment without consideration of these properties may yield poor results. Therefore the determination and consideration of these properties have an important role. HDPE (High Density Poly Ethylene) used as storage materials, since packaging of fruits in polyethylene films results in modified atmosphere which reduced the fruit decay, softening and loss soluble solids during storage. To our knowledge, detailed investigations concerning physical properties of tomato in relation with storage conditions and storage material have not been published. Therefore, the aim of this research was to see the effect on physical attributes of tomato due to HDPE as storage material at ambient storage conditions. This information provides useful insights into design of processing, packing equipments and transportations for tomato.

Materials and methods

The experiment was conducted at Food Analysis Laboratory of Sardar Vallabhbhai Patel University of Agriculture and Technology, Meerut (India). Fresh and disease free tomatoes were procured directly from the farmers of village Dhanju and Lawad. Four varieties of tomatoes namely *Badshah, Himshikhar, NS-524* and *Raja* were used for the present investigation.

Measurement of dimensions

Three linear dimensions namely polar diameter (D_1), major diameter (D_2) and minor diameter (D_3) for all tomatoes were measured using a Vernier Caliper (least count 0.01mm). Polar diameter is defined as the distance between tomato apex and the stem end. Major and minor diameters of the tomatoes are defined as maximum and minimum width respectively in a plane perpendicular to a polar axis (Mohsenin, 1986).

Mass, volume and density

Mass of fresh tomatoes was determined using high accuracy electronic balance. As the tomatoes were numbered the weight of individual tomatoes were recorded every day. The volume of tomato was determined individually by water displacement method using a cylinder of 1 liter capacity. The mass and volume were expressed in 'g' and 'ml' respectively (1 ml = 1cm3). Densities for tomatoes were calculated using the following equation:

$$\text{Density} = \frac{\text{weight(g)}}{\text{volume(cm}^3\text{)}}$$

Geometrical and morphological properties viz. AMD, GMD, surface area and sphericity; Density, Shape factor (λ) etc. were measured same as Kumar *et al.*, (2016). TSS of tomatoes was measured using a hand hold refractrometer.

Results and Discussions

In case of variety *Badshah;* dimensional and geometrical parameter like polar diameter (D_1), major diameter (D_2), minor diameter (D_3), AMD, GMD decreased with increase in storage period (4.95 to 4.93 cm, 5.20 to 5.10 cm, 5.08 to 4.95 cm, 5.075 to 4.992 cm and 5.073 to 4.990 cm respectively) (Table 1). Mass, volume, surface area and shape factor also decreased with increase in storage period (75.775 to 75.118 g, 79.50 to 66.50 ml, 81.209 to 78.576 cm^2 and 1.000 to 0.992 respectively). Whereas sphericity, density and TSS increased with increase in storage time (97.076 to 97.812 %, 1.097 to 1.138 g/cc and 9.46 to 9.58 °B respectively).

Table 1 Effect of packaging material (HDPE) and storage condition (ambient temperature) on the physical properties of tomato (variety: Badshah).

Days	D_1 (cm)	D_2 (cm)	D_3 (cm)	AMD (cm)	GMD (cm)	Sphericity (%)	Mass (g)	Volume (ml)	Surface area (cm^2)	Density (g/cc)	Shape factor	TSS (°Brix)
1	4.95 ±0.42	5.20 ±0.43	5.08 ±0.45	5.075 ±0.42	5.073 ±0.42	97.076 ±0.43	75.775 ±20.80	69.50 ±21.00	81.209 ±13.47	1.097 ±0.06	1.000 ±0.02	9.46 ±0.49
2	4.95 ±0.42	5.19 ±0.45	5.04 ±0.44	5.058 ±0.42	5.056 ±0.42	97.008 ±0.27	75.647 ±20.78	68.75 ±20.54	80.696 ±13.54	1.106 ±0.07	0.996 ±0.02	N.D.
3	4.95 ±0.42	5.18 ±0.46	5.00 ±0.43	5.042 ±0.43	5.040 ±0.43	96.939 ±0.37	75.518 ±20.77	68.00 ±20.08	80.182 ±13.63	1.116 ±0.08	0.992 ±0.02	N.D.
4	4.95 ±0.42	5.15 ±0.44	4.99 ±0.44	5.029 ±0.42	5.027 ±0.42	97.151 ±0.17	75.398 ±20.72	67.50 ±20.12	79.779 ±13.42	1.124 ±0.08	0.992 ±0.02	N.D.
5	4.95 ±0.42	5.13 ±0.42	4.98 ±0.45	5.017 ±0.41	5.015 ±0.41	97.366 ±0.59	75.278 ±20.67	67.00 ±20.17	79.374 ±13.22	1.132 ±0.09	0.992 ±0.03	N.D.
6	4.93 ±0.41	5.10 ±0.37	4.95 ±0.45	4.992 ±0.40	4.990 ±0.40	97.812 ±0.74	75.118 ±20.60	66.50 ±19.84	78.576 ±12.81	1.138 ±0.10	0.992 ±0.02	9.58 ±0.48
$CD_{5\%}$	N.S.	0.060	0.041	0.020	0.020	N.S.	0.117	1.377	0.658	0.024	N.S.	
SE(d)	0.014	0.028	0.019	0.009	0.009	0.337	0.054	0.640	0.306	0.011	0.003	
SE(m)	0.010	0.020	0.013	0.007	0.007	0.239	0.038	0.453	0.216	0.008	0.002	
CV	0.413	0.763	0.537	0.264	0.260	0.491	0.102	1.334	0.541	1.416	0.453	
R^2	0.428	0.962	0.939	0.989	0.989	0.676	0.998	0.989	0.988	0.994	0.689	

During observation of variety *Himshikhar* (Table 2), a gradual decrement was shown in mean values of all the physical parameters namely polar diameter (D_1) major diameter (D_2), minor diameter (D_3), AMD, GMD, mass, volume, sphericity, surface area, density and shape factor of tomato with increase in storage period (4.80 to 3.63 cm, 5.80 to 4.28 cm, 5.35 to 3.90 cm, 5.317 to 3.933

cm, 5.299 to 3.922 cm, 87.218 to 66.640 g, 90.00 to 64.50 ml, 91.438 to 68.873%, 88.181 to 64.422 cm², 0.973 to 0.773 g/cc and 1.011 to 0.746 respectively). Whereas only TSS increased with increase in storage time (6.550 to 6.825 °B respectively). Similar trends were reported by Varshney *et al.,* (2007).

Table 2 Effect of packaging material (HDPE) and storage condition (ambient temperature) on the physical properties of tomato (variety: Himshikhar).

Days	D_1 (cm)	D_2 (cm)	D_3 (cm)	AMD (cm)	GMD (cm)	Sphericity (%)	Mass (g)	Volume (ml)	Surface area (cm2)	Density (g/cc)	Shape factor	TSS (°Brix)
1	4.80 ±0.23	5.80 ±0.24	5.35 ±0.10	5.317 ±0.10	5.299 ±0.10	91.438 ±2.91	87.218 ±10.71	90.00 ±14.14	88.181 ±3.46	0.973 ±0.04	1.011 ±0.03	6.550 ±0.58
2	4.80 ±0.23	5.65 ±0.26	5.20 ±0.18	5.217 ±0.16	5.203 ±0.16	92.151 ±2.01	87.215 ±10.70	89.00 ±13.22	85.067 ±5.17	0.983 ±0.03	1.000 ±0.03	N.D.
3	4.79 ±0.25	5.65 ±0.26	5.20 ±0.18	5.213 ±0.16	5.198 ±0.16	92.068 ±2.06	87.200 ±10.70	88.63 ±13.28	84.912 ±5.17	0.987 ±0.03	1.001 ±0.03	N.D.
4	4.78 ±0.26	5.65 ±0.26	5.20 ±0.18	5.208 ±0.16	5.194 ±0.16	91.985 ±2.12	87.184 ±10.70	88.25 ±13.38	84.757 ±5.19	0.991 ±0.04	1.002 ±0.03	N.D.
5	4.78 ±0.26	5.63 ±0.29	5.18 ±0.21	5.192 ±0.18	5.178 ±0.18	92.114 ±2.12	86.969 ±10.83	86.75 ±11.84	84.254 ±5.84	1.004 ±0.02	1.000 ±0.03	N.D.
6	3.63 ±2.43	4.28 ±2.86	3.93 ±2.62	3.942 ±2.63	3.930 ±2.62	69.016 ±46.06	66.654 ±45.57	65.75 ±44.78	64.712 ±43.34	0.760 ±0.51	0.749 ±0.50	N.D.
$CD_{5\%}$	3.63	4.28	3.90	3.933	3.922	68.873	66.640	64.50	64.422	0.773	0.746	6.825
SE(d)	±2.43	±2.86	±2.61	±2.62	±2.62	±45.97	±45.56	±43.56	±43.10	±0.52	±0.50	±0.57
SE(m)	N.S.	N.S.	N.S.	N.S.	N.S.	N.S.	N.S.	N.S.	N.S.	N.S.	N.S.	
CV	0.791	0.953	0.877	0.873	0.870	15.775	14.103	13.070	13.795	0.186	0.174	
R^2	0.559	0.674	0.620	0.617	0.615	11.154	9.972	9.242	9.754	0.131	0.123	

In case of variety NS-524 (Table 3); decrement was observed in values of the physical parameters namely major diameter (D_2), minor diameter (D_3), AMD, GMD, mass, volume and surface area of tomato with increase in storage period (4.77 to 4.72 cm, 4.40 to 4.38 cm, 4.556 to 4.533 cm, 4.552 to 4.531 cm, 56.946 to 56.712 g, 66.67 to 62.50 ml and 65.204 to 64.596 cm² respectively). Although the samples were spoiled after four days of storage. Whereas the values of shape factor, density, sphericity and TSS increased continuously before samples get spoiled (0.966 to 0.968, 0.848 to 0.901 g/cc, 95.459 to 96.025 % and 4.933 to 5.467 °B respectively). Polar diameter (D_1) remains unchanged (4.50 cm) during storage period.

During observation of variety *Raja* (Table 4); decrement was observed in values of the physical parameter namely polar diameter (D_1), major diameter (D_2), minor diameter (D_3), AMD, GMD, mass, volume and surface area of tomato with increase in storage period (4.43 to 4.37 cm, 4.53 to 4.47 cm, 4.33 to 4.27 cm, 4.433 to 4.367 cm, 4.430 to 4.364 cm, 57.005 to 56.289 g, 56.00

to 54.33 ml and 61.835 to 60.026 cm^2 respectively). Whereas the values of density, sphericity and TSS increased continuously (1.019 to 1.037 g/cc, 95.613 to 96.232 %, 8.97 to 9.20 °B respectively). Variation occurred in shape factor during storage but the initial and final values were same (0.977).

Table 3 Effect of packaging material (HDPE) and storage condition (ambient temperature) on the physical properties of tomato (variety: NS-524).

Days	D_1 (cm)	D_2 (cm)	D_3 (cm)	AMD (cm)	GMD (cm)	Sphericity (%)	Mass (g)	Volume (ml)	Surface area (cm2)	Density (g/cc)	Shape factor	TSS (°Brix)
1	4.23 ±0.42	4.47 ±0.35	4.27 ±0.25	4.322 ±0.33	4.320 ±0.33	96.738 ±1.10	50.478 ±10.79	52.33 ±11.68	58.842 ±9.09	0.966 ±0.01	0.989 ±0.02	4.467 ±0.06
2	4.23 ±0.42	4.45 ±0.35	4.27 ±0.25	4.317 ±0.33	4.315 ±0.33	96.976 ±0.69	50.441 ±10.79	51.17 ±10.80	58.699 ±9.11	0.986 ±0.02	0.990 ±0.02	ND
3	4.23 ±0.42	4.43 ±0.35	4.27 ±0.25	4.311 ±0.34	4.310 ±0.34	97.219 ±0.31	50.405 ±10.79	50.00 ±10.00	58.555 ±9.14	1.007 ±0.04	0.991 ±0.02	ND
4	4.23 ±0.42	4.38 ±0.38	4.25 ±0.25	4.289 ±0.34	4.288 ±0.34	97.859 ±1.30	50.221 ±10.83	48.67 ±10.07	57.975 ±9.30	1.033 ±0.07	0.992 ±0.02	ND
5	Spoiled											5.067
6	Spoiled											±0.12
$CD_{5\%}$	0.396	0.341	0.240	0.321	0.320	1.046	10.281	10.298	8.722	0.052	0.021	
SE(d)	0.176	0.151	0.106	0.142	0.142	0.463	4.555	4.563	3.864	0.023	0.009	
SE(m)	0.124	0.107	0.075	0.101	0.100	0.328	3.221	3.226	2.732	0.016	0.007	
CV	7.618	6.267	4.577	6.059	6.056	0.876	16.608	16.586	12.132	4.215	1.716	
R^2	0.685	0.696	0.688	0.689	0.689	0.679	0.688	0.726	0.693	0.646	0.684	

Table 4 Effect of packaging material (HDPE) and storage condition (ambient temperature) on the physical properties of tomato (variety: Raja).

Days	D_1 (cm)	D_2 (cm)	D_3 (cm)	AMD (cm)	GMD (cm)	Sphericity (%)	Mass (g)	Volume (ml)	Surface area (cm2)	Density (g/cc)	Shape factor	TSS (°Brix)
1	4.43 ±0.23	4.53 ±0.47	4.33 ±0.38	4.433 ±0.32	4.430 ±0.32	95.613 ±3.26	57.005 ±10.26	56.00 ±10.58	61.835 ±8.79	1.019 ±0.01	0.977 ±0.02	8.97 ±0.76
2	4.43 ±0.23	4.52 ±0.45	4.32 ±0.41	4.422 ±0.33	4.419 ±0.33	95.690 ±3.12	56.880 ±10.28	55.50 ±10.04	61.539 ±8.96	1.025 ±0.02	0.976 ±0.02	N.D.
3	4.43 ±0.23	4.50 ±0.44	4.30 ±0.44	4.411 ±0.33	4.408 ±0.34	95.770 ±3.06	56.755 ±10.29	55.00 ±9.54	61.241 ±9.13	1.031 ±0.03	0.974 ±0.03	N.D.
4	4.42 ±0.25	4.50 ±0.44	4.28 ±0.42	4.400 ±0.34	4.397 ±0.34	95.895 ±2.62	56.605 ±10.28	54.67 ±9.67	60.944 ±9.16	1.035 ±0.04	0.976 ±0.02	N.D.
5	4.40 ±0.26	4.50 ±0.44	4.27 ±0.40	4.389 ±0.34	4.386 ±0.34	96.023 ±2.24	56.456 ±10.26	54.33 ±9.81	60.645 ±9.21	1.040 ±0.04	0.977 ±0.01	N.D.
6	4.37 ±0.21	4.47 ±0.40	4.27 ±0.40	4.367 ±0.32	4.364 ±0.32	96.232 ±1.43	56.289 ±10.27	54.33 ±9.81	60.026 ±8.67	1.037 ±0.04	0.977 ±0.02	9.20 ±0.75
$CD_{5\%}$	N.S.	N.S.	N.S.	0.019	0.019	N.S.	0.077	N.S.	0.533	N.S.	N.S.	

Days	D_1 (cm)	D_2 (cm)	D_3 (cm)	AMD (cm)	GMD (cm)	Sphericity (%)	Mass (g)	Volume (ml)	Surface area (cm2)	Density (g/cc)	Shape factor	TSS (°Brix)
SE(d)	0.026	0.031	0.025	0.008	0.009	0.652	0.034	0.698	0.236	0.012	0.006	
SE(m)	0.018	0.022	0.018	0.006	0.006	0.461	0.024	0.493	0.167	0.009	0.005	
CV	0.716	0.852	0.726	0.236	0.240	0.833	0.074	1.555	0.474	1.475	0.803	
R^2	0.779	0.867	0.942	0.979	0.979	0.964	0.996	0.942	0.976	0.882	0.052	

Conclusion

From the experiment it can be conclude that dimensional and geometrical parameters of all tomatoes decreased with increase in storage duration however increment was observed in values of sphericity and density for all varieties except *Himshikhar*. TSS increased gradually during storage duration for entire tomato varieties; this is due to loss of some moisture from tomatoes. Variety *NS-524* shows maximum shrinkage among all during observation period.

Abbreviations: D_1 = polar diameter (height), D_2 = major diameter, D_3 = minor diameter, GMD = geometric mean diameter, AMD = arithmetic mean diameter, TSS = total soluble solid, BOD = biological oxygen demand, CD = critical difference, SE (d) = Standard error of deviation, Se (m) = Standard error of mean, R^2 = coefficient of determination, N.D.= not detected.

References

Giovannucci, E. (1999). Tomatoes, tomato-based products, lycopene, and cancer: Review of the epidemiologic literature. Journal of the National Cancer Institute, 91(4): 317–331.

Kumar, T., Chandra, S., Samsher, Chauhan, N., Singh, J. and Arya, A. M. (2018). Storage and aluminum foil packaging dependent physical properties of tomatoes. Chem. Sci. Rev. Lett., 7(25): 113–117.

Kumar, T., Chandra, S., Singh, A. and Singh, Y. (2016). Storage and packaging dependent physical properties of tomatoes. J. Pure Appl. Microbio., 10(4): 2901–2907.

Mirzaee, E., Rafiee, S., Keyhani, A., Emam Djom-eh Z. and Kheiralipour, K. (2008). Mass modeling of two varieties of apricot (*Prunus armenaica* L.) with some physical characteristics. Plant Omics J., 1: 37–43.

Mohsenin, N.N. (1986). Physical properties of plant and animal materials. Gordon and Breach Science Publishers, pp 20–89.

National Committee on Plasticulture Applications in Horticulture (NCPAH). Ministry of Agriculture, Government of India, New Delhi. (http://www.ncpahindia.com/tomato.php).

Rao, A.V., and Agarwal, S. (1999). Role of lycopene as antioxidant carotenoid in the prevention of chronic diseases: A review. Nutrition Research, 19(2): 305–323.

Taheri-Garavand, A.; Ahmadi, H. and Gharibzahedi, S.M.T. (2009). Investigation of moisture-dependent physical and chemical properties of red lentil cultivated in Iran. International Agricultural Engineering Conference (IAEC). Bangkok, Thailand, 7–10 Dec. 2009.

Varshney, A.K., Sangani, V.P. and Antala, D.K. (2007). Effect of storage on physical and mechanical properties of tomato. Agriculture Engineering Today, 31(3–4): 47–53.

13

Engineering Properties of Garlic (*Allium sativum* L.) Bulbs and Cloves

Bogala Madhu, Vishvambhar D. Mudgal
and Padam S. Champawat

Abstract

In India, the allium groups (onion and garlic) are important bulb crops for home consumption as well as sources of income to many peasant farmers in many parts of the country. The garlic bulbs are valued for their flavor and have wide applications in food and their antimicrobial, antifungal, insecticidal and antioxidative properties. The engineering properties of garlic bulb, clove and skin plays important role in the planning of various post-harvest operations like drying, sorting, grading, bulb breaking, peeling and designing the process equipments like bulb breaker, clove separator and peeler. The aim of this chapter was to study the engineering properties viz., physical, mechanical, aerodynamic and frictional properties of garlic bulbs, and cloves in both peeled and unpeeled form.

Introduction

India is rightly the spice bowl of the world. India has the largest domestic market for spices in the world. Indian spices are the most sought-after globally, given their exquisite aroma, texture and taste. Garlic (*Allium sativum* L.) is a world's favourite, versatile horticultural commodity consumed for culinary, medicinal and antimicrobial purposes and is being cultivated for over 5000 years. The most important part of the garlic plant is the compound bulb. Each bulb is made of 6-26 cloves and is wrapped in a white papery sheath. Some varieties have a radish or purplish sheath. The bulbs are valued for their flavour, command extensive commercial importance because of their wide application in food and pharmaceutical preparations. Garlic is still probably nature's most powerful medicinal plant to us today. It is recognized to have remarkable preventive and curative abilities. Garlic is mainly used aromatic spices and pickles as it has digestive, carminative and anti-rheumatic properties. It is being used in ayurvedic formulation since a long for curing lungs, healing the intestinal ulcer and checking muscular pain and giddiness.

India is the second largest producer of garlic after China with a share of about 14 per cent of the world's area and 5.27 per cent of the world's production (FAO, 2016). In India, garlic is being grown on about 3.2 lakh hectares area with a total production of about 16.93 lakh tons. During the past 18 years, garlic consumption has increased by 45 per cent. The main losses in garlic storage are physiological weight loss (12-15%) and losses due to diseases (10 to 25%) (Tripathi *et al.,* 2009). The total losses during storage are off 25-40 per cent under ambient conditions (Tripathi and Lawande, 2006).

The physical properties of garlic such as size, shape, surface area, volume, density, porosity, colour and appearance are important in designing the graders, bulb breakers and peelers etc. Mechanical properties such as hardness, compressive strength, impact, shear resistance and rheological properties are useful in designing the bulb breakers, peelers and resistance to cracking under harvesting and handling operations. The mechanical damage to the bulb and clove in harvesting and handling causes a reduction in germination and increases chances of insect/pest infestation during storage and also affects the quality of final products. The aerodynamic properties such as terminal velocity and drag coefficient are important and required for designing the air conveying systems and separation equipments. The drag coefficient and other physical properties are required for calculating terminal velocity. The frictional properties such as coefficient of friction, angle of repose, static and sliding coefficient of friction are important for designing the storage structures, hoppers and conveying systems in the handling of garlic.

Methodology for engineering properties

Physical properties like length, width, thickness, geometric mean diameter, arithmetic mean diameter, aspect ratio, sphericity, surface area, volume, true density, bulk density and porosity of garlic bulb and cloves (peeled and unpeeled) can be determined by using the following equations as shown in Table 1 (Bakhtiari and Ahmad, 2015).

Table 1 Equations for measuring different physical properties of agricultural produce

S. No.	Physical property	Units	Equation/ Instrument
1.	Length	mm	Digital vernier calliper with an accuracy of 0.01 mm.
2.	Width	mm	
3.	Thickness	mm	
4.	Geometric Mean Diameter	mm	$(LWT)^{1/2}$
5.	Arithmetic Mean Diameter	mm^2	$\frac{L+W+T}{3}$

S. No.	Physical property	Units	Equation/ Instrument
6.	Aspect Ratio	Decimal	W/L
7.	Sphericity	Decimal	$\frac{(LWT)^{1/3}}{L}$
8.	Surface Area	mm^2	
9.	1000 kernel mass of garlic cloves	g	
10.	Volume of one clove	mm^3	Measuring cylinder
11.	True Density	Kg/m^3	Toluene displacement method using a pycnometer $\frac{\text{Mass}}{\text{Volume}}$
12.	Bulk Density	Kg/m^3	$\frac{\text{Mass}}{\text{Volume}}$
13.	Porosity	Decimal	$\left(1-\frac{\text{Bulk density}}{\text{True density}}\right)\times 100$

Angle of repose

The angle of repose for the garlic bulb and cloves (peeled and unpeeled) over stainless steel can be determined by the tilt platform method. The platform was made of Stainless steel and raised slowly. The angle when bulb/cloves start sliding/rolling was determined by tangent method and angle of repose was calculated using the following Equation.

$$\text{Angle of repose, degrees} = \text{Tan}^{-1}\left(\frac{\text{height}}{\text{base length}}\right)$$

Coefficient of static friction

The coefficient of static friction of garlic bulbs and cloves can be determined by the inclined plane method. The samples were placed on the test surface at the top edge. The inclined surface was tilted until the samples begin to move leaving an inclined surface. The angle of inclination with the horizontal was measured by a scale provided and taken as an angle of internal friction and tangent of the angle was taken as coefficient of friction between surface and sample.

Coefficient of static friction = $\tan\theta$

Where, θ = Angle of inclination of the material surface, degree

Crushing load

Crushing implies the partial or complete destruction of cloves. Clove was sat upon a flat plate until the cross-head of a handmade apparatus was brought in contact with the clove and a compression force was applied by adding weights or loads until permanent (destruction) was caused and then the loads were recorded.

Force required for loosening the cloves from the whole bulb

The force required for loosening the cloves from the whole bulb was determined by placing the whole bulb upon a flat plate until the cross-head of a handmade apparatus was brought in contact with the bulb and a compression.

Drag coefficient

The drag coefficient of the garlic cloves (*Cd*) was calculated using the following equation.

$$F_R = CA_p \times \frac{\rho_f V^2}{2}$$

Where, F_R = resistance drag force or weight of clove at terminal velocity, kg; C = Overall drag coefficient; r_f = Mass density of fluid, kg.s²/m⁴; A_p = Projected area of the clove normal to the direction of motion, m²; V = Relative velocity between the main body of fluid and material, m/s.

Terminal velocity

The suspension cloves air velocities were determined by a vertical cylinder of plexi glass which was used to observe cloves while suspended (Sahay and Singh, 1994). The air flow rate was controlled by varying the speed of the fan motor. When the samples were suspended, air velocity was recorded as terminal velocity of cloves by an anemometer to a resolution of 0.1 m/s.

$$V_t = \left[\frac{2W\,(\rho_p - \rho_f)}{\rho_p \rho_f A_p C}\right]^{1/2}$$

Where, V_t = Terminal velocity, m/s; C = Overall drag coefficient; G = Acceleration due to gravity, m/s²; ρ_p = Mass density of clove, kg.s²/m⁴; ρ_f = Mass density of fluid, kg.s²/m⁴; W = Weight of clove, kg; A_p = Projected area of the clove normal to direction of motion, m².

Engineering properties of garlic bulbs

Bahnasawy (2007) studied few selected physical, chemical and mechanical properties of the garlic. These properties like linear dimensions such as geometric mean diameter, arithmetic mean diameter, cross-sectional area were measured with calliper reading of 0.01 mm, shape index by formulae, frontal surface area by planimeter, volume and bulk density by sand displacement method, static coefficient of friction by adjustable tilting plate method, crushing load and force required to loosening the cloves from the whole bulb by texture analyzer. The results showed that the garlic geometric and arithmetic mean diameters ranged from 25.3 to 49.3 mm and 25.3 to 50.2 mm respectively according to the bulb size categories. The cloves length, width and thickness were 19.2 to 29.1, 7.8 to 13.2 and 6.9 to 9.9 mm respectively. The surface and cross-sectional of areas ranged from 5331 to 13640 and 2910 to 12840 mm^2 respectively. The number of cloves ranged from 18 to 51 / bulb according to the bulb size categories. Bulk density, angle of repose and coefficient of contact surface values ranged from 892 to 1007 kg/ m^3, 41.52 to 45.04 degree and 0.91 to 1.12 respectively, according to the bulb size categories. The mechanical properties showed that the friction angle ranged from 23.35 to 28.82 degree, where small bulbs recorded the highest values on the concrete surfaces, while the lowest values were recorded by the large bulb on the iron surfaces. The friction coefficient decreased with increasing bulb size, where it was highest (0.8) for small bulbs on the concrete surfaces: on the other hand, the lowest values (0.36) were recorded for large bulbs on iron surfaces. The crushing load of the cloves ranged from 55.6 to155 N, depending on the bulb size. The force required for loosening the cloves from the bulb ranged from 110 to 272 N and 101 to 320 N on the horizontal and vertical positions of the bulbs.

Table 2 Means of the physical properties of small, medium and large size garlic bulbs (Source: Bahnasawy, 2007)

Property	Units	Bulb size		
		Small	Medium	Large
Surface area	mm^2	5331	8289	13637
Cross-sectional of area	mm^2	2910	7210	12835
Volume	mm^3	16600	36900	84700
Mass of whole bulb	g	1672	3668	7553
Bulk Density	kg/m^3	1007	994	892
Number of Cloves per bulb		13–18	32–41	45–51

Manjunatha *et al.*, (2008) estimated the physical properties of garlic at 40.50 per cent moisture content on wet basis namely shape by comparison with the standard chart, diameter, length, width, thickness by digital micrometer to an accuracy of 0.01mm,weight by using an electronic balance of 0.001 g, bulk density by known volume container and electronic balance, specific gravity by liquid displacement method, geometric mean diameter and sphericity by formulae, angle of repose by emptying method, coefficient of static friction by calibrated tilting table, terminal velocity by an air column. The shape, average diameter, weight, bulk density, segment number of whole garlic, average length, width, thickness, geometric mean diameter, sphericity and mass weight of 1000 garlic segments as round, 51.2 cm, 28.64 gm, 414.40 kg/m^3, 16.32., 26.25 mm, 10.36 mm, 8.73 mm, 13.34 mm, 0.51 and 1813.60 g respectively. In the moisture content range from 23.05 to 40.50 per cent (w.b.), the terminal velocity, angle of repose, specific gravity, compressive and shear forces of garlic segment increased from 7.18 to 12.24 m/s, 25.53 to 37.50 degree, 0.90 to 0.97, 2.25 to 10.70 kg and 1.75 to 2.83 kg respectively, while bulk density decreased from 483.10 to 449.76 kg/m^3. The static coefficient of friction increased on three surfaces, namely, teak wood (0.46 to 0.53), aluminium (0.38 to 0.48) and mild steel (0.34 to 0.41) with an increase in moisture content from 23.05 to 40.50 per cent (w.b.).

Haciseferogullari *et al., (*2004) studied some technological properties of garlic using 10 repetitions at a moisture content of 66.32 per cent (d.b.). Technological properties such as the diameter of whole garlic, length of whole garlic, width and thickness, were measured using a micrometre to an accuracy of 0.01mm. Geometric mean diameter, sphericity and segment number of whole garlic were established. Projected areas of segments were measured by using a digital camera (Kodak DC 240) and the Sigma Scan Pro 5 program. Segment mass and thousand segments mass were measured using an electronic balance to an accuracy of 0.001g. Segment density and volume by liquid displacement method, bulk density by with hectoliter tester, porosity by porosity device, terminal velocity by air column and hardness from the forces in a Test Instrument of Biological Materials were measured using the procedure described by Aydin and Ogut (1991). The mean mass and mass weight of 1000 garlic segment, length of the segment, diameter of whole garlic, geometric mean diameter, sphericity. projected area, volume, bulk density, porosity and hardness of segments were measured as 32.81 g, 2383.8 g, 27.24 mm, 46.51 mm, 15.15 mm, 0.55, 4.54 cm^2, 2245.64 mm^3, 478.75 kg/m^3, 54.16 per cent and 13.78 N respectively. Also, static and dynamic coefficients of friction for garlic segments were established on a galvanized sheet, iron steel and plywood. These values for static and dynamic coefficients were found as 0.41-0.35, 0.47-0.40 and 0.54-0.48 respectively.

Table 3 Means of the mechanical and frictional properties of small, medium and large size garlic bulbs (*Source*: Bahnasawy, 2007)

Property		Units	Bulb size		
			Small	**Medium**	**Large**
Angle of repose		degrees	45.04	43.42	41.52
Coefficient of contact surface			0.91	1.02	1.12
Coefficient of Static Friction			0.43	0.42	0.36
Force required for loosening the garlic cloves from the whole bulb	Horizontal Position	N	110	189	272
	Vertical Position		101	181	320
Crushing load		N	55.60	83.40	155.00

Engineering properties of garlic cloves

Bakhtiari and Ahmad (2015) studied the moisture-dependent physical and aerodynamic properties of garlic cloves The average length, width, thickness, geometric and arithmetic mean diameter of garlic cloves were 32.0, 21.8, 20.9, 24.4 and 24.9 mm, respectively. The average of the surface area, projected area, one thousand kernel mass, volume and bulk density of garlic cloves increased from 1718.3 to 2029.1 mm^2, 546.6 to 644.3 mm^2, 6783.0 to 8159.3 g, 5916.5 to 7356.0 mm3 and 476.3 to 567.4 kg/m^3, respectively, with increasing moisture content from 35.8% to 60.5% w.b. Studies showed that as moisture content increased, the true density decreased from 1146.4 to 1109.3 kg/m^3. Within the same moisture range, the terminal velocity of garlic cloves increased linearly from 15.6 to 16.7 m/s. Finally, a vacuum seed metering system (a unit of pneumatic planter) for planting garlic cloves was designed and developed based on physical and aerodynamic properties of garlic cloves. Means of the physical properties of garlic cloves at different moisture contents are presented in Table 4.

Table 4 Means of the physical properties of garlic clove at different moisture contents (*Source*: Bakhtiari and Ahmad, 2015)

Property	Unit	Moisture Content (% w.b.)		
		35.8	**47.4**	**60.5**
Length	mm	30.93	31.96	33.08
Width	mm	20.93	21.64	22.83
Thickness	mm	19.89	21.08	21.73

Property	Unit	Moisture Content (% w.b.)		
		35.8	47.4	60.5
Geometric Mean Diameter	mm	23.39	24.43	25.41
Arithmetic Mean Diameter	mm^2	23.87	24.89	25.88
Aspect Ratio	Decimal	0.67	0.68	0.69
Sphericity	Decimal	0.76	0.76	0.77
Surface Area	mm^2	1718.29	1875.06	2029.13
1000 kernel mass of garlic cloves	g	6783.00	7280.67	8159.33
Volume of one clove	mm^3	5916.50	6442.70	7356.00
True Density	Kg/m^3	1146.39	1130.02	1109.28
Bulk Density	Kg/m^3	476.28	537.91	567.39
Porosity	Decimal	58.45	52.39	48.85
Terminal velocity	m/s	15.6	16.37	16.67
Drag coefficient	Decimal	0.77	0.77	0.86

Masoumi *et al.*, (2003) studied terminal velocity and frictional properties of two common types of garlic cloves (white and pink) in Iran. The static coefficient of friction against three surfaces (galvanized steel, plexiglass, and rubber)were measured using an adjustable tilting plate, emptying and filling angles of repose were measured by emptying method and terminal velocity of cloves were measured by using a hot wire anemometer at a moisture range from 34.9 to 56.7 per cent (w.b.). The maximum and minimum value of the coefficient of friction was 74 per cent for white garlic against rubber at a moisture content of 55.67 per cent (w.b.) and 26 per cent for pink garlic against galvanized steel at 35.26 per cent (w.b.) respectively. The maximum value of filling angle of repose was 43.5° for white cloves at 55.67 per cent (w.b) and minimum corresponding value was 36.1° for pink garlic cloves at 35.26 per cent (w.b.). The white garlic cloves had a minimum value of terminal velocity 9.82 m/min at 34.9 per cent (w.b.) and maximum corresponding value was 16.66 m/min at 56.7 per cent (w.b.) for pink cloves.

Masoumi *et al.*, (2006) identified and compared physical attributes of Iranian garlic cloves. A machine vision system was used to determine three dimensions and both major and minor projected areas of garlic cloves at a moisture content of 42.4 per cent (w.b.). The geometric mean diameter and sphericity were calculated as well as the unit mass and volume of cloves was measured. The moisture content of the cloves was determined by following the ASAE S352.2 standard method. Size and shape characteristics were determined by using a computer imaging system. The average physical attributes of white garlic major diameter, intermediate diameter, minor diameter, major projected area, minor projected area, unit mass and unit volume were measured as 31.71(±0.46) mm, 19.91(±0.47) mm, 15.41(±0.45) mm, 473.8(±16.1) mm^2, 377.9(±14.6) mm^2, 4.28(±0.22) g and 4401(±213) mm^3. The average physical attributes of pink garlic major diameter, intermediate diameter, minor diameter, major projected area, minor projected area, unit mass and unit volume were measured as 30.56(±0.54) mm, 18.72(±0.46) mm, 15.15(±0.41) mm, 427.6(±16.6) mm^2, 352.8(±14.15) mm2, 3.68(±0.21) g and 3809(±0.21) mm^3.

Kaur *et al.*, (2017) studied the engineering properties of peeled and unpeeled garlic cloves. The physical properties of whole garlic segments such as length, width, thickness were measured to estimate their geometric mean diameter and sphericity. Other properties such as bulk volume, bulk density, true density and porosity were also determined for both types of cloves. The average values of geometric mean diameter, sphericity, bulk density and true density for unpeeled and peeled cloves were measured as 11.820 mm, 494, 471 Kg/m^3, 1077 Kg/m^3 and 11.004 mm, 526, 556 Kg/m^3, 1169 Kg/m^3, respectively. The coefficient of external friction for unpeeled cloves for plywood and GI surface was calculated as 0.366 and 0.772 and for peeled cloves as 0.664 and 0.812, respectively. The coefficient of internal friction for the unpeeled and peeled cloves was found to be 0.570 and 0.826, while the angle of repose values was calculated as 39.88 and 23.52 degrees, respectively.

Table 5 Average values of physical properties of unpeeled and peeled cloves (*Source*: Kaur *et al.*, 2017)

Property	Unit	Unpeeled	Peeled
Average length	mm	23.822	21.002
Average width	mm	9.423	9.053
Average thickness	mm	7.331	7.168
Geometric mean diameter	mm	11.82	11.04
Sphericity	Decimal	0.490	0.526
Bulk density	Kg/m^3	471	556
True density	Kg/m^3	1077	1169
Porosity	Decimal	56.17	52.413
Coefficient of external friction		0.664 (W) 0.812 (GI)	0.366 (W) 0.772 (GI)
Coefficient of internal friction		0.570	0.826
Angle of repose	degrees	25.526	39.881
Coefficient of external friction		0.664 (W) 0.812 (GI)	0.366 (W) 0.772 (GI)

W = Wooden surface, *GI* = Galvanized iron sheet

Conclusion

The physical properties of the garlic bulb and cloves such as size, shape, surface area, volume, density, porosity, colour and appearance are important in designing the graders, bulb breakers and peelers etc. Mechanical properties such as hardness, compressive strength, impact, shear resistance and rheological properties are useful in designing the bulb breakers, peelers and resistance to cracking under harvesting and handling operations. The aerodynamic properties such as terminal velocity and drag coefficient are important and required for designing the air conveying systems and separation equipments. The drag coefficient and other physical properties are required for calculating terminal velocity. The frictional properties such as coefficient of friction, angle of repose, static and sliding coefficient of friction are important for designing the storage structures, hoppers and conveying systems in the handling of garlic.

References

Aydın, C. and Öğüt, H. (1991). Determination of some biological properties of Amasya apple and hazelnuts. Selcuk Univ J Agric Fac; 1(1):45–54.

Bahnasawy, A.H. (2007). Studies on some physical and mechanical properties of garlic. International Journal of Food Engineering, 3(6):1–18.

Bakhtiari, M.R. and Ahmad, D. (2015). Determining physical and aerodynamic properties of garlic to design and develop of pneumatic garlic clove metering system. Agricultural Engineering International: CIGR Journal, 17(1): 59–67.

Haciseferogullari, H., Musa, O., Fikret, D. and Sedat, C. (2004). Some nutritional and technological properties of garlic. Journal of Food Engineering, 68: 463–469.

Kaur, M., Kaur, P., and Kaur, A. (2017). Engineering Properties of Peeled and Unpeeled Garlic Cloves. Agricultural Research Journal, 54 (1): 85–89.

Manjunatha, M., Samuel, D.V.K. and Jha, S.K. (2008). Studies on some engineering properties of garlic (Allium sativum). Journal of Agricultural Engineering, 45(2):18–23.

Masoumi, A.A., Rajabipoor, A., Tabil, L.G. and Akram, A.A. (2006). Physical Attributes of Garlic (*Allium sativum* L.). Journal of Agricultural Sciences and Technology, 8: 15–23.

Masoumi, A.A., Rajabipour, A., Tabil, L. and Akram, A A. (2003). Paper presented in the meeting of The Canadian society for engineering in agricultural, food, and biological systems. Paper NO.03-330:1–9.

Sahay, K.M. and Singh, K.K. (1994). Unit Operations of Agricultural Processing. Vikas Publishing House Pvt. Ltd., New Delhi.

Tripathi, P.C and Lawande, K.E. (2006). Cold Storage of Onion and Garlic. Technical Bulletin No. 15.

Tripathi, P.C, Sankar, V. and Lawande, K.E. (2009). Effect of storage environment and packing methods on storage losses in garlic. Indian Journal of Horticulture, 66: 511–515.

14

Physical Properties of Foods

Upendra Singh, Amit Kumar and S.K. Goyal

Abstract

Physical properties of food are aspects such as colour, structure, texture, rheology and interfacial properties, and composition. We have a range of instrumental methods for objectively characterising and measuring food structure and physical properties. These are useful for applications such as new product development, benchmarking, reformulation and specification.

Colour

Consistent and accurate measurements of the colour and visual appearance of food products is extremely important. Various methods are available for colour measurement, allowing a wide variety of sample types to be measured. Colour measurement results are typically provided on the CIELAB scale. Others are available on request.

Structure

The structure of food influences texture. Examples include porous products such as aerated foods and bakery products where the bubble structure affects softness, and starch-based snacks where it affects crispiness.

Food structure analysis using X-ray micro-CT

X-ray micro-CT offers non-destructive imaging and structure measurement in 3D. Images and movies showing the internal structure of products can be generated. Measurements of porosity, bubble size distribution and structure thickness (wall size) can be performed.

Texture

Food texture is an important sensory attribute as it affects the way food tastes and how it feels in the mouth. The texture depends on the rheological properties of the food and evaluation involves measuring the response of a food when it is subjected to forces such as cutting, shearing, chewing, compressing or stretching.

Rheology and interfacial properties

The rheological properties of food materials are important in determining the texture as well as how they behave physically when subjected to physical forces and forced to flow. The rheological properties of raw materials, intermediate products such as batters and doughs as well as final products can be studied.

Thermal analysis

Thermal analysis techniques measure the physical and chemical properties of foods as a function of temperature or time.

Compositional mapping

Many food products have a non uniform distribution of composition. For example, fried products have a higher fat content near surfaces, and baked products have a higher moisture content in the centre of the product. Compositional mapping techniques allow these gradients to be measured and visualised.

Grain quality

In agriculture, grain quality depends on the use of the grain. In ethanol production, the chemical composition of grain such as starch contents is important, in food and feed manufacturing, properties such as protein, oil and sugar are significant, in milling industry soundness is the most important factor to consider and for seed producer, the high germination percentage (viability of seed) and seed dormancy is the important feature to consider, for consumers the properties like colour and flavour will be important.

Properties of grain quality

Overall quality of grain is affected by several factors includes, growing practices, time and type of harvesting, postharvest handling, storage management and transportation practices. The properties of grain quality can be summarized into ten main factors (i) Uniform moisture contents, (ii) High test weight, (iii) No foreign material, (iv) Low percentage of discoloured, broken and damaged kernels, (v) Low breakability, (vi) High milling quality, (vii) High protein and oil content, (viii) High viability, (ix) No afaltoxin (mycotoxin), and (x) No presence of insects and molds.

Characteristics of grain quality

Grain quality is characterized into two main factors (i) intrinsic factors, and (ii) extrinsic factors. The intrinsic factors of grain includes, colour, composition, bulk density, odour, aroma, size and shape. Colour is an important primary factor for characterization and grading, trade, and processing of grain. It is a common criterion used in wheat trade. The main compositions of grain are carbohydrates (energy), protein, lipids, mineral, fiber, phytic acid, and tannins. It varies significant depends on the type of grain, genetics, varieties, agricultural practice, and handling of the grain. Grain composition plays a significant role in grading and marketing of grains. Bulk density is defined as the ratio of the mass to a given volume of a grain sample including the interstitial voids between the particles. Size and shape are important factors in grain quality and grading; it varies between grain to grain and between varieties of the same species. It is commonly used in rice grading and key factors in milling industry. The extrinsic factors include: age, broken grain, immature grain, foreign matter, infected grain and moisture content.

Grain quality and grade specification

Grain grading and specification system assures that a particular lot of grain meets the required set standards customer. In many countries grading of grain depends on four main properties; (i) bushel (test) weight (ii) moisture contents (iii) broken foreign material or the percentage fragments example broken corn foreign materials (iv) damaged kernels (i.e. total and heat damaged).

Moisture contents: The moisture content is one among important factors in grains quality. It denotes as the quality of water per unit mass of grain and expressed on a percentage basis (i.e. wet basis or dry basis). Moisture content does not directly affect grain quality but can indirectly affect quality since grain will spoil at moisture contents above that recommended for storage.

Foreign material (FM): Broken foreign material is an important factor in grading and classification of grains. It is described as foreign material other than grains such as sands, pieces of rocks, plastics particles, metals and pieces of glass, contaminating a particular lot of grain. In the grains trade presences of more than set percentage of FM results either low grades, price discount or lot rejection, because the higher the FM the more the cost to clean before uses.

Damaged kernel (DK): Damaged kernels constitute an important grading factor. DK are considered those that have an evident visual damaged and negatively affect their value of the grains. It is usually quantified by removing damaged kernels by hand from potions free of impurities. Grain grades contain a limit of damage kernels for each grade, for instance for wheat to be grade one must contain no more than 0.4% of the total weight. Main types of damaged are due to insects, heat, molds, weathering, sprouted, frost, diseases, non-uniform maturity and lack of/partial grain filling. In the grading systems or specification damaged kernels is divided into two main parts (i) heat damaged and (ii) total damaged.

Non-grain-standards properties

Important non-grain standards in U.S. grain standards includes (i) breakage susceptibility, (ii) milling quality (iii) seed viability, (iv) nutritive value, (v) mold count and carcinogen content, and (vi) insect infestation and damage.

Best example of grain quality can be described into two common grains (wheat and rice).

Wheat

Wheat grain (*Triticum aestivum* L.) is the world's leading agricultural source of energy, protein and fiber; it belongs to a family Graminaea and genus Triticum and can be categorized into three main classes - hard, soft and durum. Wheat quality can best be described in terms of end-user, nutritional quality, milling, and baking and rheology quality. In general wheat needs to be sound, clean, well mature, free from foreign material and damaged.

Wheat quality

In general wheat quality can be divided into three main groups (i) botanical (species and varieties), (ii) physical (iii) and chemical characteristics.

Botanical criteria of quality in botanical terms, wheat quality can be described as falling into the following two main criteria (i) species and (ii) varieties.

Physical criteria: Physical characteristics of wheat quality includes, grain weight, hardness, grain size, shape, vitreousness and colour. Physical properties of grain such as wheat play a very important role in the quality of the grain, and in final products such as flour. Main physical properties that influence quality of wheat are test weight, hardness, grain size & shape, vitreousness and colour.

Test weight: Test weight of wheat is considered the most common and easiest way to quantify wheat. It is an important quality factor in wheat grading as it gives rough estimates of flour yields. The basic factors that affect the test weight of wheat are kernel size and shape, kernel density, maturity of wheat, anseases and actual wheat variety.

Hardness: The hardness of wheat endosperm is critical in determining the suitability of wheat for various end products and influences the processing and milling of wheat. It is the common characteristic used by millers and trader to classify wheat. In term of hardness wheat can be classified as either hard or soft.

Colour: In terms of colour, wheat is classified into two classes (i.e. red wheat and white wheat), hard red winter wheat is considered superior and commonly used for bread flour production, while white wheat are usually used for cake, chapattis, and pasta (macaroni), each types of wheat has different properties such as taste, baking quality and milling yields.

Vitreousness: Wheat vitreousness is an optical property used by many countries to grade or quantify durum wheat. Based on vitreouness, wheat can be classified into three main classes: vitreous, mealy and piebald. Vitreous wheat differs from non-vitreous by kernel appearance (starchy and opaque); vitreous wheat are considered better quality than non-vitreous kernels, because of higher quality semolina protein, nice colour and uniform coarser granulation.

Chemical (quality) properties of wheat: A chemical property of wheat includes moisture contents, protein (gluten) contents, amylase content and fibre contents.

Moisture contents: Wheat grain normal harvested at 10-12% moisture contents. In most countries moisture contents is not part of grading system, but is the most important factors affecting quality of wheat grain, hence is inversely related to dry matter loss. Moisture contents has two significance important in wheat quality (i) too low (too dry) result wheat to break during storage and handling operation and (ii) too high will facilitate molds growth which lead to deterioration.

Protein content: Protein is not part of wheat a grading factors, but its quantity and quality are the most important properties in wheat business. Most buyers and millers need to know the amount of protein contents of wheat before buys it. Wheat contains five different classes of protein; (i) albumin (soluble in water), (ii) globulin (soluble in salt solution) (iii) gliadin (soluble in 70% aqueous ethanol) (iv) proteose and (v) glutenin (soluble in dilute acid or alkali).

Other important quality of wheat are milling and baking quality are as follows:

Milling quality: Most of wheat is commercially sold as milled flour or semolina, hence milling quality is a crucial factor in wheat trade. Milling depends on three main factors (i) size and evenness of kernels- there is a close correlation with the weight of grain, determined by thousand-kernel weight, (ii) texture of the endosperm- characterized by glassiness or pearling index and hardness. They influence the utilization of energy required for milling as well as the amount of semolina obtained, and (iii) percentage ratio of the seed-coat- the larger the kernel the lower the ratio of seed-coat, and if the layers are not thicker, then the percentage of the seed-coat will decreases too, and colour of endosperm and seed-coat.

Baking quality: Baking quality is another criterion used to determine the quality and suitability of wheat; baking quality depends on types of wheat uses and processing conditions, for instance the strong (hard) wheat are considered of the higher quality and suitable for bread making, where most of cakes made from soft wheat flour. Baking quality is determined by rheological properties of wheat flour. The rheological property of wheat flour is essential because it determine other physical characteristics such as dough (baking) volume and sensory attributes.

Grading and classification of wheat quality: Wheat, like other cereal, is graded based on certain criteria such as; test weight, purity, maximum percentage damaged and foreign materials. In the United States, wheat is classified into classes and sub-classes. In classes, wheat is split into eight different groups; hard red spring, hard red winter, soft red winter, durum, hard white, soft white, mixed and un-classed wheat. These classes are further sub-classed into five grades (US. No.1-5) with the exception of un-classed wheat.

Rice

Rice belongs to the genus Oryza of the sub-tribe Oryzinea in the family Gramineae. Three main categories of rice are, (i) long-grain- relatively long and bold types, known as Carolina rice, (ii) medium-grain-long, thin, cylindrical grain, known as Patna, and (iii) short-grain-short, stout grain, known as Spanish-Japan.

Rice grain quality

There is no proper definition or description of rice quality, because as definition of quality, depends on several factors such cooking practice and region and usages for example rice miller he/she describe rice quality in terms of total recovery and or head and broken rice kernels while food processing will define concept of rice quality in terms of grain size, aroma, appearance and cook ability.

In general many countries quantify rice into four main categories (i) milling quality (ii) cooking, eating and processing quality, (iii) nutritional quality and (iv) specific standards for cleanliness, soundness and purity. In the United States three more factors has been added (i) hull and pericarp (ii) colour grain size, shape, weight, uniformity and general appearance and (iii) kernel chalkiness, translucency and colour.

Physical properties of rice: Common physical properties of rice are size, shape, colour, uniformity, and general appearance. Other factors contributes to general appearance of rice are cleanliness, free from other seeds, vitreousness, translucency, chalkiness, colour, damaged and imperfect kernel.

For the case of grain size, rice grain can be categorised into three main groups (i) length (ii) shape and (iii) weight, the length is the measure of the rough, brown, or milled rice kernel in its highest dimension, while shape is the ratio of length, width, and thickness, and for the case of weight is determined by using 1,000-kernel weight.

Test weight: Test weight is another important grading factor of rice, it is related to bulk density, and used to measure the relative amount of foreign material or immature kernels; it is useful index in milling outturn. The average test weight per bushel of U.S. rough rice is 45 lb.

Impurities and damaged rice: Impurities and damaged rice considered as single most important factor of rice quality because it is directly related to economic value of lot, example presence of sand and stones will increase the weight of grain and damage rubber when send to the miller. Impurities and damaged rice contains dockage, damaged kernels, chalkly grains, red rice, broken seed or kernels and odors.

Milling quality or out-turn: The main objective of rice milling is to remove the outer layer (hull), bran and germ with minimum damage of endosperm. Milling quality of rice is another important criteria used in marketing, grading and classification of rice, as well as treatment such as conditioning, drying and parboiling, it is normally estimated by using milling yield. Milling yield varied depends on several factors such as grain types, varieties, chalkiness, drying and storing conditions, other includes environmental conditions and moisture contents at harvest.

The milling quality can be determined by two common parameters (i) total yield and (ii) head yield, also another parameters like degree of milling and broken rice are used to estimate milling quality an express in percentage. By definition milling quality is the ability of rice kernels to stand milling and or polishing without breakage, and to yield higher amount of recovery.

Quality factor in foods

In countries where food is abundant, people choose foods based on a number of factors which can in sum be thought of as quality. Quality has been defined as degree of excellence and included such things as taste, appearance, and nutritional content. We might also say that quality is the composite of characteristics that have significance and make for acceptability. Acceptability, however, can be highly subjective. Quality and price need not go together, but food manufacturers know that they generally can get a higher price for or can sell a larger quantity of products with superior quality. Often "value" is through of as a composite of cost and quality. More expensive foods can be good value if their quality is similar for all practical purpose, yet the price can very as much as threefold depending on other attributes of quality. This is why processors will go to extremes to control quality.

When we select foods and when we eat, we use all of our physical senses, including sight, touch, smell, taste, and even hearing. The snap of a potato chip, the crackle of a breakfast cereal, and crunch of celery are textural characteristics, but we also hear them. Food quality detectable by our senses can be divided into three main categories: appearance factors, textural factors, and flavour factors. Appearance factors include such things as size, shape, wholeness, and different forms of damage, gloss, transparency, colour, and consistency. For example, apple juice is sold both as cloudy and clear juice. Each has a different appearance and is often thought of as a somewhat different product. Textural factors include hand feel and mouth feel of firmness, softness, juiciness, chewiness, grittiness. The texture of a food is often a major determinant of how little or well we like a food. For example, many people do not like cooked liver because of its texture. Texture of foods can be measured with sophisticated mechanical testing machines. Flavour factors include both sensations perceived by the tongue which include sweet, salty, sour, and bitter, and aromas perceived by the nose. The former are often referred to as "flavours" and the latter "aromas," although these terms are often used interchangeably. Flavour and aroma are often subjective, difficult to measure accurately, and difficult to get a group of people to agree. A part of food science called sensory science is dedicated to finding ways to use humans to accurately describe the flavors and other sensory properties of foods. There are hundreds of descriptive terms that have been invented to describe flavour, depending on the type of food. Expert tea tasters have a language all of their own, which has been passed down to members of their guild from generation to generation. This is true of wine tasters as well.

Quality of a food product involves maintenance or improvement of the key attributes of the product—including colour, flavour, texture, safety, healthfulness, shelf life, and convenience. To maintain quality, it is important to control microbiological spoilage, enzymatic degradation, and chemical degradation. These components of quality depend upon the composition of the food, processing methods, packaging, and storage.

Appearance Factors

Of the sensory attributes of food, those related to appearance are the most susceptible to objective measurement, but appearance is important to the consumers. They have certain expectations of how food should look. Two separate categories of appearance include:

- Colour attributes
- Geometric attributes (size and shape)

Colour

Of these two, colour is by far the most important. Consumers expect meat to be red, apple juice to be light brown and clear, orange juice to be orange, egg yolks to be bright yellow-orange, and so on. Food colour measurements provide an objective index of food quality. Colour is an indication of ripeness or spoilage. The end point of cooking processes is judged by colour. Changes in expected colours can also indicate problems with the processing or packaging. Browns and blackish colours can be either enzymatic or non enzymatic reactions.

Maillard reaction

This is the dominant browning reaction. Other less explained reactions include blackening in potatoes or the browning in orange juice. The enzymatic browning found widespread in fruits and selected vegetables is due to the enzymatic catalyzed oxidation.

Phenolic compounds

Naturally occurring Pigments play a role in food colour. Water-soluble pigments may be categorized as anthocyanins and anthoxthanins. Lesser known water-soluble pigments include the leucoanthocyanins. Fat-soluble plant pigments are primarily categorized into the chlorophyll and carotenoid pigments. These greenand orange-yellow pigments considerably impact the colour. Myoglobins contribute to the colour of meat.

Measuring Colour

In order to maintain quality, the colour of food products must be measured and standardized. If a food is transparent, like a juice or a coloured extract, colorimeters or spectrophotometers can be used for colour measurement. The colour of liquid or solid foods can be measured by comparing their reflected colour to defined (standardized) colour tiles or chips. For a further measurement of colour, reflected light from a food can be divided into three components:. The color of a food can be precisely defined with numbers for these three components with tri-stimulus colorimetric. Instruments such as the Hunter lab Colour and Colour Difference Meter measure the value, hue, and chrome of foods for comparisons.

Size and Shape

Depending on the product, consumers expect foods to have certain sizes and shapes. For example, consumers have some idea of what an ideal French fry should look like, or an apple, or a cookie, or a pickle. Size and shape are easily measured. Fruits and vegetables are graded based on their size and shape, and this is done by the openings they will pass through during grading. Now computerized electronic equipment can determine the size and shape of foods.

Flavour Factors

Food flavour includes taste sensations perceived by the tongue sweet, salty, sour, and bitter—and smells perceived by the nose. Often the terms flavour and smell (aroma) are used interchange-ably. Food flavour and aroma are difficult to measure and difficult to get people to agree on. A part of food science called sensory science is dedicated to finding ways to help humans accurately describe the flavours and other sensory properties of their food. Flavour, like colour and texture, is a quality factor. It influences the decision to purchase and to consume a food product. Food flavour is a combination of taste and smell, and it is very subjective and difficult to measure. People differ in their ability to detect tastes and odors. People also differ in their preferences for these. Besides the tastes of sweet, salty, sour, and bitter, an endless number of compounds give food characteristic aromas.

Resources

Bern, C., and T. J. Brumm. (2009). Grain Test Weight Deception. Iowa State University-University Extension. PMR 1005,

Brooker, D.B., F.W.Bakker-Arkem, and C.W.Hall. (1992). Drying and Storage of Grains and Oilseeds. An AVI Book, Van Nostrand Reinhold, New York.

Cornell, H, J., and Hovelling, A. W. (1998). Wheat: Chemistry and Utilization. Technomic Publishing Company, Inc., Lancaster, Pennsylvania.

Corriher, S. O. (1997). Cookwise: The hows and whys of successful cooking. New York: William Morrow and Company, Inc.Cremer, M. L.

Dowell, F.E. (2000). Differentiating vitreous and non-vitreous durum wheat kernels by using near-infrared spectroscopy. Cereal Chem. 77(2): 155–158.

Gooding, M.J., and Davies, W.P. (1997). Wheat Production and Utilization, Systems, Quality and the Environment. Cab International., New York USA.

Henry. R. J., and P.S Kettlewell. (1996). Cereal grain quality (Ist edi.). Chapman and Hall. London UK.

Houston, D.F. (1972). RICE: Chemistry and Technology (edited), American Association of Cereal Chemists, St. Paul, Minnesota.

Ktenioudaki, A., Butler, F. and Gallagher, E. (2010). Rheological properties and baking quality of wheat varieties from various geographical regions. Journal of Cereal Science, 51: 402–408.

Li, Y., Shoemaker, F.C., Maa, J., Moon, K.J. and Zhong, F. (2008). Structure-viscosity relationships for starches from different rice varieties during heating. Food Chemistry, 106:1105–1112.

Pomeranz, Y. (1964). Wheat Chemistry and Technology (edited). American Association of Cereal Chemists. St. Paul, Minnesota.

Pomeranz, Y. (1987). Modern Cereal Science and Technology. New York. VCH Publishers, Inc.

Ram, M.S., Dowell, F.E., Seitz, L. and Lookhart. G. (2002). Development of standard procedures for a simple, rapid test to determine wheat color class. Cereal Chem. 79(2): 230–237.

Roy, P., Ijiri, T., Okadome, H., Nei, D., Orikasa, T., Nakamura, N. and Shiina, T. (2008). Effect of processing conditions on overall energy consumption and quality of rice (*Oryza sativa* L.). Journal of Food Engineering, 89: 343–348.

Samson, M.F., Mabille, F., Chéret, R., Abécassis, J. and More, M.H. (2005). Mechanical and physicochemical characterization of vitreous and mealy durum wheat endosperm. Cereal Chem. 82(1): 81–87.

Serna-Saldivar, S.O. (2012). Cereal grains Laboratory Reference and Procedures Manual. Food Preservation Technology Series. LLC NW. CRC Press. Taylor and Francis Group.

Singh, R.K., Singh, U.S. and Khush, G.S. (2000). Aromatic Rice. (ed). Science Publishers, Inc., New Hampshire USA.

USDA (2009). United States Department of Agriculture. United States Standards for Rice. Federal Grain Inspection Service.

Vaclavik, V. A., and E. W. Christina. (1999). On food and cooking. The science and lore of the kitchen. Simon and Schuster Inc., New York

Wrigley, C.W. (1995). Identification of Food-Grain Varieties (Edited). American Association of Cereal Chemists, St. Paul. MN USA. AACC. Inc.

15

Rheological Properties of Cereal Starch, Dough and Flours

Aneena E.R. and Simla Thomas

Humankind has always been a perceptive feel for rheological testing, e.g. in physical and visual evaluations of material properties such as hardness, stiffness, flexibility, and viscosity, and their relation to end-use quality characteristics. People often naturally measure the quality of solid foods by gently squeezing them, or liquid viscosity is measured by gently rotating the liquid in its container. These intuitive measurements gradually became formalized into quantitative descriptions of material properties.

Rheology can be defined as the study of how materials deform, flow or fail when force is applied. The name is derivated from Greek word: rheos, meaning the river, flowing, streaming. Therefore rheology means "flow science". Rheological investigations not only include flow behaviour of liquids, but also deformation behaviour of solids. Normally, to measure rheological properties, the material is subjected to a controlled, précised and quantifiable distortion or strain over a given time and the material parameters such as stiffness, modulus, viscosity, hardness, strength or toughness are determined by considering the subsequent forces or stresses (Dobraszczyk and Morgenstern, 2003). The term rheology was coined by Eugene C. Bingham, a professor at Lafayette College, in 1920, from a suggestion by a colleague, Markus Reiner (Steffe, 1996)

Food rheology focuses on the flow properties of single food components, which might already display a complex rheological response function, the flow of a composite food matrix, and the effect of processing on the food structure and its properties. For processed food the composition and the addition of ingredients to obtain a certain food quality and product performance requires deep rheological understanding of single ingredients their relation to food processing, and their final discernment (Fischer and Windhab, 2011).

The rheological measurements help to get a quantitative description of the material's mechanical properties, helps to gain information related to the molecular structure and composition of the material and also to characterize and guess the material's performance during processing and for quality control (Dobraszczyk, 2003).

Rheological measurements are an important tool to aid in process control and process design, it tells us how dough will behave under a given set of conditions and can be used to describe and guess its performance during practical processing (like during mixing, sheeting and baking of dough). Moreover, it can also be related to product functionality. Many rheological tests are used to predict end product quality such as mixing behaviour, sheeting and baking performance.

In order to examine product performance and consumer acceptance, rheological instrumentation and measurements have become essential tools in analytical laboratories. For predicting storage, stability measurements, understanding and designing texture, knowledge of the rheological and mechanical properties of different food systems is important. Rheological properties should be independent of size, shape and how they are measured; in other words, they are worldwide, rather like the speed of light or density of water, which do not depend on how much light or water is being measured or how it is being measured. In short, the rheological approach is that the properties that measured are reproducible and can be compared between different samples, test sizes and shapes, and test methods.

Cereals, over time, represented a great importance to human kind and it covered the world's largest planted area of all crops. The need for the crushing operations in the food industry results from the fact that the cereals grains are rarely used in their original shape and size. Rheological properties of the cereal grains serve as parameters for engineers designing the grain processing technologies and grinding techniques.

Rheological properties of cereal starch

Starch and proteins are the main macromolecules in cereal grains. Starch properties arise from the unique behavior of its components in aqueous systems. Starch is composed of two types of a-D-glucose polymers: amylose, which is the lower molecular linear component, and amylopectin, which is both large in size and highly branched. Amylose accounts for 20-35% of normal starch, depending upon the botanical origin. In normal cereal starches, this content is remarkably constant at 27-28%. When dispersed in cold water, starch undergoes a limited swelling, but since absorption of water is low (~30%), there is no discernible rheological effect. With sufficient heat a starch-water system or a starch-containing food material undergoes a series of dramatic changes referred to as gelatinization and pasting. These changes occur beyond the gelatinization temperature (60°C in starch systems containing excess water). Gelatinization temperature is not a constant, but greatly depends upon the characteristics (water content, dissolved solutes) of the medium. It is the changes that occur after gelatinization that result in the development of interesting rheological properties.

In the characterization of rheological property of starch, viscosity is an important parameter because starch is often utilized as a thickener in different applications. Viscosity measures the resistance of a fluid or semi-fluid to flow when a shear stress is applied. A native-starch suspension even at a high concentration (35–40%, w/w), display a low viscosity at the ambient temperature. After heating to above the gelatinization temperature, starch granules lose crystalline structure, absorb water, swell, some disperse, and develop significant viscosity. This process of viscosity development is known as starch pasting. The viscosity of the resultant starch paste determines the thickening power of starch for various applications. After cooling and storage, some starch pastes (e.g., normal wheat and maize) at an adequate concentration (greater than or equal to 6%, w/w) can form gels with a defined shape without fluidity, reflecting the concept that starch molecules within granules form networks to immobilize water and exhibit a viscoelastic property. Some starch pastes (e.g., waxy maize, tapioca, and potato), however, remain as pastes or form rather weak gels at a higher concentration. Rheological characteristics of starch paste and gel depend on many factors, including the chemical structure of starch, starch concentration, pasting conditions (e.g., temperature, shear rate, and heating rate), and storage conditions (temperature and time).

Pasting property

Pasting properties of starch can be measured using an amylograph, such as Brabender Amylograph and Rapid Visco-Analyzer (RVA) or using a dynamic rheometer in a flow temperature ramp mode. Waxy maize and waxy rice starches display lower pasting-temperatures (69.5 and 64.1°C, respectively) but higher peak-viscosities [205 and 205 Rapid Viscosity Unit (RVU), respectively] than the normal maize and rice starches (82.0 and 79.9°C, 152 and 113 RVU, respectively). Amylopectin is the primary component of starch responsible for the swelling power and viscosity development of starch during cooking, whereas amylose, particularly with the presence of lipids, tends to intertwine with amylopectin and restrict the swelling of starch granules. During cooling, amylose interacts with other starch molecules and forms networks, which substantially contributes to the setback viscosity. Consequently, normal maize and rice starches exhibit greater setback-viscosities (74 and 64 RVU, respectively) than their waxy counterparts (16 and 16 RVU, respectively). Minor components of starch granules, such as lipids and phosphate-monoester derivatives, remarkably affect the pasting property. ALC formed in starch during cooking renders entanglements with amylopectin molecules and restricts the swelling of granules, which results in a higher pasting-temperature and a lower peak-viscosity. Wheat and barley starches consist of larger amounts of phospholipids, which readily complex with amylose. Consequently, these two starches exhibit higher pasting-temperatures and lower peak-viscosities than the other normal cereal starches. When the endogenous lipids of wheat starch are removed using a detergent (e.g., sodium dodecyl sulfate), the starch displays a pasting temperature and peak viscosity similar to tapioca and waxy maize starch. In contrast, the phosphate-monoester derivatives of potato starch carry negative charges, repel one another, and enhance the swelling of starch granules, which result in a substantially lower pasting-temperature (63.5 °C) and higher peak-viscosity (702 RVU).The remarkably great peak-viscosity of potato starch is also contributed by its large granule sizes (diameter up to 75mm). Addition of sugars, including sucrose, glucose, fructose, maltose, galactose, and lactose, increases the viscosity of starch, which is attributed to the water-binding ability of the sugars. At a low concentration (< 1.0 %, w/w) salts display minimal effects on the pasting property of most native starch except potato starch. Salts substantially decrease the viscosity of potato starch because the cations of salts mask the negative charges of the phosphate-monoester derivatives and reduce the charge repulsion. Addition of lipids reduces the viscosity of normal starch and produces a short paste because of the ALC formation as discussed earlier, but little effect is observed for waxy starch because of lacking amylose.

Viscosity

Viscosity of a starch paste can be measured using a viscometer, including capillary flow, orifice, falling ball, and rotational type, or using an amylograph. Viscosity of a starch paste usually displays a non-Newtonian feature: the shear stress does not increase linearly with the increase in shear rate. The shear stress of starch paste can be expressed as a function of shear rate by fitting them into different models, such as power law, Herschel–Bulkley, and Bingham model. Viscosity of a starch paste is also thixotropic (shear thinning): It exhibits a decreased viscosity with respect to shear rate and time. Also, viscosity of a starch paste increases with starch concentration but decreases with starch amylose content and determination temperature.

Gel formation

Different methods have been used to characterize the rheological property of a starch gel. The most frequently used methods include: (1) determination of starch gel strength using a texture analyzer, which provides a “single-point” measurement; (2) dynamic modulus analysis of starch gel using a dynamic rheometer, which allows continuous assessment of starch gel at various temperatures and shear rates. A dynamic rheometer can evaluate the storage modulus (G'), the loss modulus (G"), and the loss tangent (tan d ¼ G"/G') of a starch gel. G' measures the deformation energy recovered per cycle of deformation, representing the elastic behavior of the gel; G" measures the energy dissipated as heat per cycle of deformation, representing the viscous behavior. A small tan d (G' is much larger than G") indicates that the deformation is essentially recoverable and the starch gel is stiff, behaving more like a solid; whereas a large tan d (G' is much smaller than G") reflects that the energy used to deform the gel is dissipated viscously and the starch gel is less stiff, behaving more like a liquid. The formation of a starch gel from a paste is a result of the interactions between amylose and amylopectin molecules in the granules and the formation of networks to hold water in the swollen granules. The short-term development of starch gel strength or stiffness after cooking is primarily a result of amylose gelation. Native starch with a greater amylose-content tends to develop a stronger gel at a faster rate. The strength or stiffness of a starch gel continues to increase during storage, which is a result of the recrystallization of amylopectin. Swollen starch granules with integrity, which can fill up a container, are essential for the formation of a strong gel Normal maize and pea starches can form a strong gel at 6–8% (w/w) concentration, whereas waxy maize, tapioca, and potato starches fail to form a gel at the same concentration. The differences can be ascribed to the fact that the granules of waxymaize, tapioca, and potato starch readily swell and disperse during cooking because of lacking ALC formation to maintain the integrity of swollen granules. Addition of soylecithin, oleic acid, or linoleic acid (10%, w/w, dry starch basis) to tapioca starch before cooking facilitates the gel formation at 8% (w/w) starch concentration because the formed ALC leads to controlled swelling of the starch granules and maintains the integrity. Addition of sugars, including sucrose, glucose, fructose, and maltose, in general, reduces the gel strength and G' of sago starch (6% starch, w/v). The effects can be attributed to restricted granule swelling resulting from water binding with the sugars. Salt solutions (0.5 M) that increase the gelatinization temperature and DH of starch, including Na_2SO_4, $MgCl_2$, $CaCl_2$, NaCl, and KCl, increase the gel strength of sago starch (6% starch, w/v). Salt solutions that decrease the gelatinization temperature and DH of starch, including NaI, NaSCN, KI, and KSCN, however, decrease the gel strength of sago starch (6% starch, w/v). Mainly, Na_2SO_4, $MgCl_2$, $CaCl_2$, NaCl, and KCl increase the water structure and stabilize the starch granules at a low concentration,which favors the formation of a strong gel; whereas NaI, NaSCN, KI, and KSCN break the water structure, destabilize starch granules, and cause dispersion of starch granules, which impedes the gel formation.

Rheological property of modified starch

Especially chemically and physically modified starches, acid-thinned starch exhibits a reduced viscosity because of the depolymerisation of starch molecules. Acid-thinned starch prepared using very mild acid-hydrolysis displays increased gel strength or stiffness, which is ascribed to the feature that limited acid-hydrolysis releases more linear starch molecules for the gelation process. Starch obtained after prolonged acid-hydrolysis produces a weak gel because of reduced molecular-weights. Rheological properties of oxidized and cross-linked starch depend on the level of modification. Lightly oxidized starch (e.g. <2% active chlorine concentration)shows an increase in the peak viscosity (except potato starch), which is attributed to the repulsion between the introduced carboxyl groups and the cross-linking with the aldehyde groups. Highly oxidized starch (e.g., 2–5% active chlorine concentration) displays a decreased viscosity because of the depolymerisation of starch molecules. Oxidized cassava and barley starches display lower gel-strength than the respective control starches, resulting from the depolymerisation of starch molecules and charge repulsion. Lightly cross-linked starch displays an increased peak viscosity because of increased starch molecular-weights. Highly cross-linked starch, however, shows an increased pasting-temperature and a decreased viscosity because the extensive cross-linking inhibits the swelling of starch granules. The paste of cross-linked starch has improved stability towards thermal and mechanical processing. Therefore, cross-linked starch is a preferred choice of thickeners and stabilizers used in food systems. Cross-linked potato starch (using 80–500 ppm $POCl_3$) forms a gel with a larger G' and a smaller tan d than the control starch, suggesting improved gelling-ability of the modified starch. But the gel-strength results of other starches cross-linked by using 100 ppm $POCl_3$ show inconsistent trends compared with that of the control starch. The discrepant results suggest that different starches need different levels of cross-linking for optimal functions. Substitution of starch with chemical derivatives, in general, renders the swelling of starch granules. Consequently, the substituted starch displays a lower pasting temperature and a higher peak-viscosity than the control starch, and the difference is more significant for the chemical derivatives with charges [e.g., cationic, phosphate (monoester), carboxymethyl, and octenyl succinic groups]. If the derivatization reaction causes severe disruption of starch granules and/or the derivatized chemical groups carry charges, the modified starch can develop viscosity in cold water. For example, carboxymethylated starch with a high degree of substitution (DS) readily develops viscosity in cold water and forms a clear solution. Other substituted starches, however, show the opposite pasting profile. For example, acetylated starch with a DS greater than 0.7 displays a much lower peak-viscosity than the control starch, which is attributed to two factors: (1) the loss of granular structure after the high DS modification; (2) the esterification with acetyl groups increases the hydrophobicity of the starch and reduces the hydration capacity. Compared with the control starch, substituted-starch pastes display inconsistent changes in theviscosity, depending on the modification method, DS, and starch origin. Because chemical derivatization reactions disrupt the granular structure of starch and cause great granular swelling, and the derivatives interfere with the network formation between starch chains, the substituted starch usually forms a weaker gel than the control starch.

The gel strength of cationic high-amylose maize starch increases with an increased DS because the modified starch has a decreased gelatinization temperature and can be gelatinized and swell at the boiling-water temperature, which is favorable for the gel formation. Annealed or HM-treated starch shows an increased pasting-temperature and a decreased peak- and breakdown viscosity because annealing and HMT enhance associations between starch molecules. Annealed or HM-treated starches show varied paste viscosities compared with the control starch. Because annealing and HMT enhance molecular associations and restrict the swelling of starch granules, the modified starch tends to preserve the integrity of swollen granules after cooking. Consequently, annealed or HM-treated starch generally forms a firmer gel than the control starch, and the increasing effect appears to be more obvious for native starch having a large swelling capacity, such as potato starch. But HMT has also been reported to impede the gel formation of rice and sweet potato starch. HHP-treated starches do not show a clear trend of changes inthe pasting properties compared with the control starch, and the results are dependent on the starch source and treatment condition. HHP-treated (8.6% starch suspension, 600 MPa, 15 min, 20–25 °C) normal wheat, maize, and pea starches form weaker gels than the control starch, resulting from the fact that the HHPT destroys the granular integrity and reduces the swelling of starch. Similar to acid-thinned starch, gamma radiation, electron beam, UV light, or microwave treated starch displays a decreased viscosity of the paste because of the depolymerisation of starch molecules. These results reflect the depolymerisation and cross-linking effects of radiation treatment on starch granules. Starches hydrolyzed by a-amylase from Bacillus amyloliquefaciens, b-amylase from Bacillus cereus, orisoamylase from Pseudomonas amyloderamosa exhibit decreased viscosities in comparison with the control starch. Normal rice starch hydrolyzed by a-amylase from Bacillus licheniformis fails to form a gel because of the depolymerisation of starch molecules.

Rheological properties of cereal dough

Rheology is one of the valuable tool that gives a quantitative measure for the amount of stress in the dough, which is closely related to the quality of the molecular gluten network. Rheological measurements on dough are used to define its physical properties. Dough refers to a wet mass developed after mixing of wheat flour, water and other ingredients. Physico-chemical properties of dough play important role in the bakery, pasta and ready-to-eat cereals processing industry. Dough is developed due to complex interactions among wheat constituents during mixing operation. The process of dough development begins with addition of water and commencement of mixing operation. Initially all ingredients are hydrated and appear like a sticky paste. On further mixing, the viscosity increases, sticky characteristics of dough disappear and a non-sticky mass is developed at peak consistency of dough. At this stage the dough behaves like a viscoelastic mass with both elastic and extensible characteristics.

The role of energy during mixing is crucial in the development of dough. In the process of developing dough particularly bread dough the aim is to bring about changes in the physical properties of the dough to improve its ability to retain the carbon dioxide gas produced during yeast fermentation. This improvement in gas retention ability is particularly important when the dough pieces reach the baking oven. In the early stages of baking the yeast activity is at its greatest and large volume of carbon dioxide gas is generated and released from solution in the aqueous phase of the dough. The dough pieces expand enormously at this time and well –developed dough or gluten structure is essential to withstand the internal pressure of the expanding dough.

Hydration of flour results in the formation of a visco-elastic dough (ie; both elastic and extensible). The rheology of dough is attributed to:

Gluten proteins (i.e. gliadin and glutennin): The long chain glutennin chain has extensive sites for cross linking and therefore contribute mainly to dough elasticity. Gliadin molecule which are smaller and more symmetrical have less surface area available for bonding with other protein molecule. As bonding is less between the spherical gliadin molecule than between glutenin molecules, there is limited capacity for gliadin to crosslink with neibouring molecules (ie; intermolecular bonding) and gliadin contribute less to elasticity and more to the extensibility of the dough.

Bonding (intra and intermolecular) covalent bonds (30–100 kcal/mol) disulphide bonds which form between cysteine and cysteine amino acids in the protein chain contribute to dough elasticity. The thiol-disulphide interchange reaction has been proposed to explain the extensible (i.e.; viscous) characteristic of the dough. In the first stage of this reaction between protein molecules a naturally present thiol compound opens the 1,2disulphide bond. The confirmation of the protein molecule then changes as a result of Brownian motion. In the second stage of this reaction a new disulphide crosslink forms between positions 2 and 3 on the neighbouring protein surfaces. The net effect of these stages is an increase in extensibility of the dough as the dough has irreversibly changed shape. Ionic bonds (10-20kcal/mol) in theory ions may enhance association and dissociation of dough components. In practice, the former prevails, effect of the addition of salt to the dough formulation on dough elasticity or rigidity. Ionic bonds increase dough rigidity. Hydrogen bonds (weak links: 2-10kcal/mol) hydrogen bond between small molecules. Water increases the plasticity of dough. In macro molecules hydrogen bonds may lead to elastic or even rigid structures. Hydrophobic bonds (very weak bonds: 1–3 kcal/mol) these bonds occur between the hydrophobic side groups of protein chains and contribute to elasticity and plasticity.

Rheological properties of cereal flour

Rheological studies are one of the most convenient methods for measuring indicators of quality and texture of food products. Cereal starches possess a wide application in the food industry as an additive but the native starch is not widely used in food industry due to its poor functional properties. Modified starches and flours have become important in processed foods because the functional properties of the starches and flours are improved over the native starches and cereal flours.Heat-moisture treatment (HMT; heating at a restricted moisture level) is one method used to physically modify starches and cereal flours.

High-temperature extrusion cooking is used extensively by many food industries to produce various food products with unique texture and flavor characteristics. Desirable properties in the end product are obtained by varying the processing conditions as well as the composition of the raw material. It is recognized that the addition of ingredients such as lipids, proteins, sugar, and salt alter the physical and chemical properties of the extruded foods. Changes in the properties of starchy foods caused by the addition of lipids are attributed to the formation of complexes between amylose and lipids. Twin screw extrusion studies have been conducted with cereal flours and grits that contain protein, which also binds lipids.

Differential scanning calorimetry (DSC) has commonly been used for investigating thermal behavior of starches. Gelatinization as well as retrogradation are two important thermal behaviors. According to Krueger, (1987) DSC data can provide a quantitative measure of gelatinization retrogradation and the phase transitions on maize and wheat starches, as well as on rice. Therefore, an endothermic transition could be observed by DSC. It was indicated that rice starch was thermally degraded and starch structural changes were caused by high-pressure steaming.

A rapid visco analyzer (RVA) model starch Master (Newport Scientific, Warrie Wood, Australia) was used to determine the pasting properties of blends using the following procedure: Switch on the RVA and allow it to warm up for 30 min prior to the experiment. Weigh 3.0 g (14% moisture basis) of flour in a canister. Place the paddle into the canister and vigorously jog the blade through the sample up and down 10 times or until it mixes uniformly. Insert the canister into the pre-adjusted instrument. Initiate the measurement cycle by depressing the motor tower of the instrument. Remove the canister on completion of test and discard.

Thermal properties was done by using DSC (Mettler). Heating rate was 5°C/min. from 20–150°C. Samples (110–120 mg) were mixed with water (1:2, w/w) and kept at 4°C overnight to allow a uniform distribution of water in the flours. Samples were sealed in stainless crucibles and reweighed before the DSC analysis. Onset temperature (To), peak temperature (Tp), end set temperature (Te), and enthalpy change (ΔH) were assessed. Degree of gelatinization It was estimated by using a DSC. It was calculated as follows:

$$DG\ (\%) = \frac{\Delta H_{\text{raw}} - \Delta h_{\text{sample}}}{\Delta H_{\text{raw}}}$$

Swelling Power

Swelling power was measured using a method reported by (Tester and Morrison, 1994). One-half of a gram of sample was dispersed in 15 mL of distilled water. The suspension was heated at 90°C in a water bath for 30 min with vigorous shaking. The starch gel was then centrifuged at 3000 rpm for 15 min. the weight of sediment was used to calculate the swelling power. Swelling power was calculated as follow:

$$\text{Swelling power } (\%) = \frac{\text{Mass of swollen sample (g)}}{\text{Initial mass of sample (g)}} \times 100$$

Native rice flour had maximum swelling power 8.65%. Temperature has a significant effect on swelling power as starch granules swell more on increasing temperature. This may be because of higher absorption of starch on increasing temperature. Wheat flour showed least swelling as compared to other cereal flours because of the reason that amylose-amylopectin ratio is less in wheat flour. Lawal *et al.*, (2005) reported an increase in swelling and solubility indices in oxidized and thinned hybrid maize starch which was a result of increased mobility of starch molecules, which facilitated easy percolation of water.

Viscosity

Viscosity was measured using Brookfield viscometer at room temperature with a spindle no. 3 at a speed of 30 rpm. Slurry was prepared by mixing 50 g of modified flour with distilled water to make a volume up to 500 mL.

Significant differences were obtained in the viscosity among native and extruded cereal flours. Raw cereal flours were more viscous than their relative processed cereal flours. Native rice flour had maximum viscosity (3521 cp) relative to other flours. Wheat flour had least viscosity (1154 cp) because of the difference in amylose-amylopectin ratio, which is more in wheat as compared to rice. On modification of the cereal flours, mean viscosity reduced and followed the order rice flour (1991 cp) > flour in combination (1853 cp) > wheat flour (828 cp), irrespective to the extrusion variables. Increase in barrel temperature from 175 to 190°C decreased the flour viscosity. High temperature caused starch degradation, thereby lowering the viscosity. Flour processed at 190°C had significant lower viscosity relative to the flour processed at 175°C.

Pasting Properties of Cereal Flours

The pasting properties of a material reflect its structure. During gelatinization, starch granules swell to several times to their initial volume. Swelling is accompanied by leaching of granule constituents; predominately amylose and the formation of a three dimensional network (Eliasson, 1985). These changes are responsible for the pasting characteristics exhibited by starch suspensions during heating and shearing. The pasting behavior of raw and modified cereal flours was measured using a RVA. Significant difference ($p \leq 0.05$) was observed among the pasting properties of native and high temperature short time (HTST) modified cereal flours. Temperature required for the flour paste to attain the maximum viscosity was 77.2°C (rice flour), 94.8°C (wheat flour), and 90.2°C (flour in combination). Peak viscosity occurs at the equilibrium point between swelling and polymer leaching which caused an increase in viscosity. Peak viscosity for unmodified rice flours was 2338 and final viscosity for different cereal flours differed significantly for unprocessed and processed cereal flours. The lower final viscosity of HTST processed cereal flours was the result of starch liquification during processing at higher temperature.When the starch is cooled in water, its cohesive forces within the swollen granule get weakened, and the viscosity of the paste gets decreased as the integrity of the granule is lost. The peak did not form in the dextrinized cereal starch and crosslinked corn starch. The reason of this might be that dextrinized modified flour had very low viscosity. These results were similar to the findings of Luallen (1985) who reported that dextrin had a very low viscosity and very high dissociation value. Lai, (2001) documented that peak viscosity, final viscosity, and breakdown viscosity of treated rice flour were lesser than that of raw rice flour. During setback, cooling occurs and reassociation between starch molecules, especially amylose resulted in formulation of gel structure and, therefore, viscosity increased during this phase. This phase is generally related to retrogradation and reordering of starch granules. Native cereal flours showed higher reordering of starch granules in the order rice flour (1821 cp) > wheat (1485 cp) > flour in combination (1534 cp). Flours processed at higher temperature (190°C) were less viscous than the flours extruded at 175°C.

References

Dobraszczyk, B.J. (2003). Measuring the Rheological Properties of Dough. In: Breadmaking Improving Quality. Woodhead Publishing. Cambridge, UK. pp. 375–400.

Dobraszczyk, B.J. and Morgenstern M.P. (2003). Rheology and the bread making process. J. Cereal Sci., 38(3):229–245.

Eliasson, A.C. (1985). Starch in Food. Woodhead Publishing Limited: London, UK.

Fischer, P. and Windhab, E.J. 92011). Rheology of food materials. Current Opinion in Colloid and Interface. Science., 16: 36–40.

Krueger, B.R., Walker, C.E., Knutson, C.A. and Inglett, G.E. (1987). Differential Scanning Calorimetry of Raw and Annealed Starch Isolated from Normal and Mutant Maize Genotypes. Cereal Chemistry, 54: 187–189

Lai, H.M. (2001). Effects of Hydrothermal Treatment on the Physicochemical Properties of Pregelatinized Rice Flour. Food Chemistry, 72:, 455–463

Lawal, O.S., Adebowale, K.O., Ogrinsanwo, B.M., Barba, L.L. (2005). N.S. Oxidized and Acid Thinned Starch Derivatives of Hybrid Maize: Functional Characteristics, Wide Angle X Ray Diffractometry, and Thermal Properties. International Journal of Biology and Macromology, 35:, 71–79

Luallen, T.E. (1985). Starch As a Functional Ingredient. Food Technology, 39: 59–63. 27.

Steffe. J.F. (1996). Rheological Methods in Food Process Engineering. Freeman Press. ISBN 978-0-9632036-1-8.

Tester, R.F. and Morrison, W.R. (1994). Properties of Damaged Starch Granules. V. Composition and Swelling of Fractions of Wheat Starch in Water at Various Temperatures. Journal of Cereal Science, 20: 175–181.

16

Texture and Viscosity Measurements

Insha Zahoor and Mohammad Ali Khan

Abstract

Food texture is one of the most widely measured quality attributes during postharvest handling, processing, and consumption. Given the subjectivity in human perception of food texture, texture measurement remains a complex exercise and thus presents both a charm and challenge for researchers and industry practitioners. For fresh foods such as fruit and vegetable, textural properties such as firmness are widely used as indices of readiness to harvest (maturity) to meet requirements for long term handling, storage and acceptability by the consumer. For processed foods, understanding texture properties is important for the control of processing operations such as heating, frying and drying to attain desired quality attributes of the end product. Texture and viscosity measurement is one of the most common techniques and procedures in food and postharvest research and industrial practice. Foods are complex systems exhibiting various degrees of elasticity, viscosity, and plasticity. Because of its complex structure and mechanical behavior, objective measurements of food texture may be influenced by a variety of test conditions, including rate of loading, the magnitude of deformations imposed upon the material, geometry of the loading surface, and localized yielding within the product tested. Various approaches have been used to evaluate the sensory attributes of texture in foods. However, the high cost and time consumption of organizing panelists and preparing food limit their use, and often, sensory texture evaluation is applied in combination with instrumental measurement.

Introduction

Fruits and vegetable products exist in various physical forms from simple Newtonian fluids (clarified juices), disperse systems (juices and pulps) to solids (fresh and processed fruits and vegetables). Fluids, however viscous yield a time to stress and begin to flow, but solids, however plastic, require a certain magnitude of stress before they begin to flow. The science of rheology is devoted to the study of flow and deformation.

Texture is one of the key quality attributes used in the fresh and processed food industry to assess product quality and acceptability. Texture attributes are also used along the food value chain to monitor and control quality, ranging from decision about readiness to harvest to assessing the impacts of postharvest handling and processing operation on product shelf life and consumer preference and acceptability. Postharvest handling and operating conditions such as storage temperature usually have distinct effects on food texture properties.

Textural quality attributes of food may be evaluated by descriptive sensory (subjective) or instrumental (objective) analyses. The combination of time and high cost associated with sensory analysis has motivated the development and widespread use of empirical mechanical tests which correlate with sensory analysis of food texture. Over the years, a wide range of instrumental tests has been used in both research and industry to assess food texture. Often the choice of any particular instrument and analytical procedure depends on costs and availability of expertise within the organization.

Principles of texture measurements

Most texture measurements are based on force measurements, but other principles such as distance, time, energy, and miscellaneous tests may be used (Bourne, 1990).

- The *puncture* principle involves measuring the force required to push a probe into a food. This principle is widely used for fruits, vegetables, and gels. The researcher believes that the technique has the potential for wider use on baked goods.
- The *extrusion* principle involves applying force to a food until it flows through an outlet that is usually in the form of one or more slots or holes. This test principle is seldom used on finished baked goods, although it has been used for frostings and fillings.
- *Cutting shear* involves cutting across a piece of food. This test principle is often mistakenly called a shear test. It is often used for meat and meat products.
- *Gentle compression* is the application of a small nondestructive force to measure deformability. There are two ways of performing this test: (1) measure the force required to achieve a standard compression (the American Association of Cereal Chemists Bread Firmness Test uses this principle), and (2) measure the distance the product deforms under a standard force.
- The *crushing* principle involves subjecting the food to a high degree of compression until it breaks up. This destructive test is the basis of the texture

profile analysis procedure (TPA), which will be described more fully below.

- The *tensile* principle involves the force required to break a food in tension. Although it has been occasionally used for evaluating bread and dough, it is infrequently used with other foods.
- *Texture profile analysis* is a special kind of test that should be set apart from the single-point tests described above. The TPA procedure was developed by a group at the General Foods Corporation Technical Center. Most researchers now use a universal testing machine to perform TPA, and many use a digitizer interfaced to a computer or direct computer readout of the data

This test consists of compressing a bite-size piece of food two times in a reciprocating motion that imitates the action of the jaw and extracting from the resulting force-time curve a number of textural parameters that correlate well with sensory evaluation of those parameters. Fig. 1 shows a generalized texture profile analysis curve obtained in the Instron Universal Testing Machine. Since this test simulates mastication, the degree of compression should be high; 80–90% compression levels are usually used in this highly destructive test.

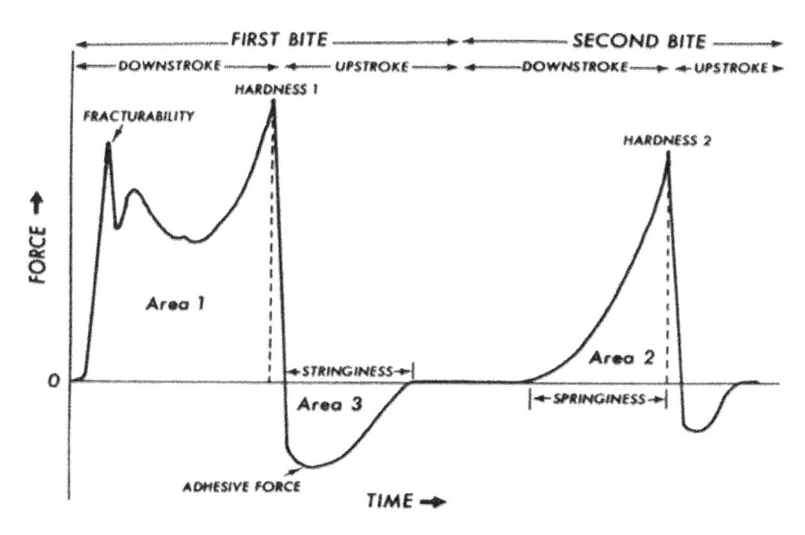

Fig. 1 A generalized texture profile analysis curve showing the parameters that are measured. The ratio: Area 2/Area 1 give the cohesiveness (Adopted from Bourne, 1990)

Selection of a suitable test principle

In view of the large number of test principles and instruments that can be used for measuring the texture of food, a cereal chemist can easily become bewildered when faced with the problem of developing a suitable procedure for measuring the textural properties of baked goods. The procedure outlined below is designed to guide one through the necessary steps to select the correct test principle for each particular application in the shortest possible time (Bourne, 1982).

Step 1: Determine the nature of the product. The kind of material (crisp, aerated, homogeneous, plastic, brittle, heterogeneous) affects the type of test principle that should be used. For example, the extrusion test is unsuited for baked goods, because baked goods do not flow under force. Viscosity measuring instruments cannot be used on baked goods, but they can be used on doughs and precooked cereals made into baby food.

Step 2: Define the purpose of the test. Will the test be used for quality control, for product development, for setting official legal standards, or for basic research? The answer to this question is an integral part of the selection process.

Step 3: Determine the degree of accuracy required. A large sample size or a greater number of replicate tests gives a higher degree of accuracy, but this approach requires more product, a higher force range, and more time to perform the test. A compromise has to be made between the cost and time of the test and the degree of accuracy required

Step 4: Decide if the test is to be destructive or nondestructive. Destructive tests ruin the structure of the sample, rendering it unsuitable for repeating the tests or for using the product for other purposes. Nondestructive tests leave the food in a condition close to its original state, so the test can be repeated on the same item. Both types of tests are used in the food industry. Because mastication is a highly destructive process, it seems logical that destructive tests should be the predominant type of test to be used on foods.

Step 5: Figure the costs. The cost of a test includes the initial cost of the instrument and its maintenance and operating costs, and the labor costs for the operator of the instrument. A simple instrument can be operated by unskilled or semiskilled personnel, whereas sophisticated instruments need to be operated by a person with higher qualifications. Although the initial cost is higher, automated readout and calculations from the data are often more economical in the long run because of labor savings and the reduced risk of errors.

Step 6: Consider time. The amount of time that can be spent on the test should be taken into account. Routine quality control tests require an instrument that gives results rapidly, while some basic research requires sophisticated tests and the amount of time required is not of great consequence.

Step 7: Decide on a location. Where will the instrument be used? An instrument may need to withstand heat, dust, vibration, and other hazards on the plant floor. Some instruments cannot withstand such an environment. All instruments can be used inside a clean, quiet laboratory.

Measurement of food texture

Since texture perceived in the mouth largely depends on the mechanical behavior of food which will determine the dynamics of breakdown during eating, most of the objective measurement research is based on the mechanical/rheological properties of foods. A wide range of destructive and non-destructive methodologies and relevant instruments has been used to measure the texture of fresh and processed foods (Chen and Opara, 2013).

Destructive methods

Three-point bending test: During the three-point bending test, force is applied to the center of the sample, such as biscuits, potato crisps and cornflakes, by an anvil until fracture occurs. The cross head speed ranges from 1 mm/min to 120 mm/min. Based on the data of fracture stress and strain, the Young's modulus of food material can be achieved (Kim *et al.*, 2012).

Single-edge notched bend (SENB) test: The SENB test is a well-established test method, in which, the test specimens have to satisfy the standard requirements for their geometry. Similar to the three-point bending test, the whole instrumental test strip is placed across two support anvils. The notch is made on the underside, and force is applied from the top to the center of the test strip by a third anvil until fracture occurs. The speed of the third anvil is often at 2 mm/s. Fracture toughness (including critical stress intensity factor and fracture energy) of food, for example biscuits or apples, can be evaluated.

Compression and puncture test: Compression test and puncture test are the most common methods to measure food texture properties. The testing foods may be solid or semi-solid. For instance, the compressing sample foods include gels, gram, apple rings, cornflakes, cheese, cellular cornstarch extrudates, bread crumb and carrot; the puncturing sample foods include apple, kiwifruit, potato slices and cereal snacks. These types of texture tests can be carried out on 'whole fruit' or 'parts' (skin, pulp or skin and pulp) depending on the research purpose. Sometimes, the two tests have been employed in one research. For both compression and puncture tests, the probes are usually cylindrical in shape, while the diameters of heads are quite different. In compression tests, they can be 10 mm, 25 mm, 80 mm and even 150 mm like a plate. In puncture tests, the diameter of the head (plunger) is often smaller such as 11 mm, 2 mm, or even 1 mm like a needle. With regard to the puncture cross head testing speed, it is likely to be several millimeters per second such as 4 mm/s; however, the compression cross head speed can range from 10 mm/min to 30 mm/s. Similarly, the puncture depth is usually several millimeters mainly depending on the size of samples while deformation of the original height by compressing can be 75%, 50% and 25% depending on the mechanical properties of samples as long as the maximum compression force is achieved. Based on literature evidence, the minimum number of samples for each measurement is five. In these compression or puncture experiments, the performing force, deformation percentage or puncture depth and cross head speed are the concerned parameters. However, the relationships between the force or other property parameters and the cross-head speed are seldom discussed.

Texture profile analysis (TPA) test, which is based on the imitation of mastication or chewing process, is performed with double-compression cycles. For irregular shape testing, the food sample is often cut into cylindrical shapes. For example, Jaworska and Bernas (2010) cut cylindrical samples (20 mm in length and diameter) out of mushrooms (caps and stipes, respectively). Through TPA test, a wide range of food texture properties, such as hardness, springiness, cohesiveness, adhesiveness, resiliency, fracturability, wateriness, gumminess, sliminess, and chewiness, can be analyzed.

Magness-Taylor puncture test (M-T) is the current industry standard method for fruit flesh firmness analysis, which is based on force-deformation characteristics of the fruit flesh mimicking the "mouth-feeling" of the consumer. In this test, the maximum force is recorded as a measure of fruit firmness.

Stress relaxation test: Stress relaxation test is mainly used to analyze the viscoelastic property of semi-solid foods, such as fish, cheese, sausage and flour dough. During measurement, the sample is compressed to an expected strain at a certain speed and the decreasing force is recorded during the relaxation time which may range from 1 min to 10 min. The decaying stress and the applied stain are related by relaxation modulus, where the relaxation modulus is usually calculated by using a generalized Maxwell model.

Warner–Bratzler shear force (WBSF) test: The WBSF test has been used since the 1930s and remains the most widely used instrumental measure of meat tenderness. The head or blade can be mounted on different texture analyzing machines such as the Texture Analyzer, Instron devices or other universal test machines. During testing, meat sample cores are sheared perpendicular to the muscle fiber orientation. A minimum of six 1.27 cm cores from throughout the steak, and an instrument crosshead speed of 200–250 mm/min are required following the guidelines of the American Meat Science Association (AMSA, 1995). Usually, the most considered parameter of the force–distance/ time curve is the maximum shear force.

Tests using a combination of mechanical and acoustic methods

Typical characteristic for many hard, crispy and crunchy solid products is their brittle fracture behavior, mostly accompanied by a sharp sound (acoustic emission or vibration) which is closely related to their texture attributes. Thus, researchers have combined mechanical tests, such as compression, penetration or three-point bending test, with acoustic signal analysis. For both dry–crispy and wet–crispy foods, the methods of evaluation can be divided into two groups, namely mechanical devices combined with acoustic emission detector (AED) and those combined with piezoelectric sensor. When used in combination with the AED test, force–displacement and sound amplitude–time signals were simultaneously recorded and the results showed that major acoustic signals were observed together with application of force. This coincidence was interpreted as an energy release in the form of sound as a result of material fracturing. From the sound data, the maximum sound pressure, number of sound peaks, sound curve length and area under amplitude–time curve were obtained. Products perceived as uncrispy emitted signals with lower average amplitude and higher peaks, at low frequencies and opposed a high mechanical resistance to compression. On the other hand, the crispiest flakes emitted sounds with larger average amplitude and fewer high peaks, uniformly distributed in the frequency domain with a moderated mechanical resistance. This mechanical–acoustic combining strategy has been successfully applied to measure crispness of fruits such as apple which was demonstrated to correlate with human sensory perception. Recently, an increasing number of researchers have adopted the Texture Analyzer (TA-XT plus) in texture measurement since the AED has become one selective part of the instrument. Other instruments combining a mechanical device with a piezoelectric sensor have been used to detect the vibration produced by fracture when a probe is inserted into a food product. The types of food products studied using this method include potato chips, cabbage leaves, pear, persimmon and grape flesh.

Imitative methods

Destructive methods which mimic the biting process during eating follow the motion of the bite by incisors or mastication by molars and are often referred to as "tooth methods". For example, Varela *et al.,* (2009) used tooth-like probes to compress snacks, which proved to be as good as traditional penetration tests to assess crispy characteristics. It was also demonstrated that results obtained at slow and high tooth-like probe speeds could be complementary, showing the parameters obtained at lower test speeds to be better correlated to human perception and the in-mouth fracture pattern to be more effectively characterized at higher compression speeds. Lately, Chung *et al.*, (2012) developed a method to characterize the texture attributes of semi-solid foods during "instrumental mastication", where artificial saliva can be used. Experiments showed that this technique can be used to monitor textural changes of starch-based food products during oral processing.

Other destructive methods

Except for the methods discussed above, there still are some other useful destructive tests, for example, probe tensile separation method which is employed to measure the stickiness of fluid foods, cutting-shear test which is used to evaluate the degree of cells being held together and the cutting force of fresh food, and the traction test, which has been proved to be a valid technique that can be used to measure fundamental mechanical parameters of food during a certain period.

Non-destructive methods

Non-destructive testing of texture in fresh and processed foods is critical for monitoring and controlling product quality.

Mechanical techniques

Non-destructive mechanical techniques used in food texture include the measurement of quasi-static force-deformation, impact response, "finger" compression, and bioyield detection. Similar to the destructive measurement, the mechanical methods are often combined with the indirect ones. For example, the product is excited by means of a small impact, and the vibration (about 20–20,000 Hz) is measured using a microphone, piezoelectric sensors or laser vibrometers.

Measurement of viscosity

Viscosity is defined as the internal friction of a fluid or its tendency to resist flow. Both gases and liquids have viscosity but viscosity of gases will not be discussed because there are no gaseous foods. However, some foods contain entrained gases. For example, ice cream is typically 50% air by volume, and apple flesh may contain 25% gas by volume. Some highly extruded crispy snack foods such as corn curls exceed 90% air by volume. Jones *et al.*, (2000) showed that in 36 branded ready-to-eat breakfast cereals the volume attributed to pores ranged from 68.2% for flakes made from a mixture of corn, wheat, oats and barley to 99.5% for puffed wheat. At first sight the distinction between texture and viscosity seems simple – texture applies to solid foods and viscosity applies to fluid foods. Unfortunately, the distinction between solids and liquids is so blurred that it is impossible to clearly demarcate between texture and viscosity. While rock candy can definitely be considered as a solid and milk a liquid, there are many solid foods that exhibit some of the properties of liquids and many liquid foods that exhibit some of the properties of solids. Some apparently solid foods behave like liquids when sufficient stress is applied.

Some important definitions in viscometry are set out below.

Laminar Flow and Turbulent Flow: Laminar flow is streamline flow in a fluid. Turbulent flow is fluid flow in which the velocity varies erratically in magnitude and direction.

Dynamic Viscosity: Dynamic viscosity which is frequently called 'viscosity,' or 'absolute viscosity,' is the internal friction of a liquid or its tendency to resist flow. It is usually denoted by η. According to the International Organization for Standardization (ISO) the unit of measurement for dynamic viscosity is the Pascal second. Since the Pascal second is a large unit of measurement, a more common unit for low viscosity fluids is the millipascal second.

Fluidity: It is the reciprocal of dynamic viscosity. It is occasionally used in place of viscosity. It is denoted by f.

Kinematic Viscosity: This is defined as the absolute viscosity divided by the density of the fluid. The SI unit for kinematic viscosity is the meter-square-second. An obsolete unit of kinematic viscosity is the stoke (after Stokes, 1819– 1903). One centistoke equals 0.01 stoke.

Kinematic viscosity is measured in efflux viscometers because the rate of flow in this type of viscometer is proportional to density as well as viscosity.

Kinematic viscosity is widely used in the petroleum industry where the specific gravity of liquid hydrocarbons does not vary widely. Kinematic viscosity is not used in the food industry to the same extent because a wide range of densities can be encountered, which compresses the kinematic viscosity into a smaller range than the absolute viscosity.

Relative Viscosity: This is sometimes called the 'viscosity ratio,' which is the ratio of the viscosity of a solution to the viscosity of the pure solvent.

Apparent Viscosity: This is the viscosity of a non-Newtonian fluid expressed as though it were a Newtonian fluid. It is a coefficient calculated from empirical data as if the fluid obeyed Newton's law. The symbol η_a is used to denote apparent viscosity.

Shear Stress: Shear stress is the stress component applied tangential to the plane on which the force acts. It is expressed in units of force per unit area. It is a force vector that possesses both magnitude and direction. The SI unit for shear stress is the Pascal (Pa) with units of newton meter^{-2}.

Shear Rate: Shear rate is the velocity gradient established in a fluid as a result of an applied shear stress. It is expressed in units of reciprocal seconds.

Types of Viscous Behavior

Newtonian Fluids: This is true viscous flow. The shear rate is directly proportional to the shear stress and the viscosity is independent of the shear rate within the laminar flow range. Typical Newtonian fluids are water, and watery beverages such as tea, coffee, beer, and carbonated beverages, sugar syrups, most honeys, edible oils, filtered juices, and milk. A Newtonian fluid possesses the simplest type of flow properties. A fluid with high viscosity is called 'viscous' whereas a fluid with low viscosity is called 'mobile.'

Non-Newtonian Fluids: Most fluid and semi-fluid foods fall into one of several classes of non- Newtonian fluids. Fig. 2 shows the shear stress versus shear rate plots for various types of flow.

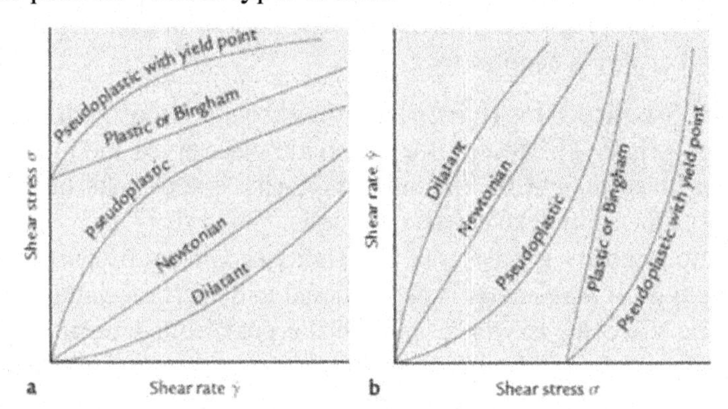

Fig 2 (a) Shear stress versus shear rate plots for various types of flow; (b) Shear rate versus shear stress plots for the same types of flow (Adopted from Bourne, 2002).

Plastic (or Bingham): A minimum shear stress known as the 'yield stress' must be exceeded before flow begins. This type of flow is often found in foods. Typical examples of this type of flow are tomato catsup, mayonnaise, whipped cream, whipped egg white, and margarine. This type of flow is named after Bingham (1922), who studied the flow properties of printing inks and discovered the important principle that no flow occurs at low stress. He identified the point at which flow begins as the 'yield stress.' The term 'plastic' refers to materials that exhibit this yield stress; it does not refer to synthetic plastics.

Pseudoplastic: In this type of flow an increasing shear force gives a more than proportional increase in shear rate, but the curve begins at the origin. The term 'pseudoplastic' was originated by Williamson (1929); it does not refer to synthetic plastics. Salad dressings are a good example of this type of flow. Many pseudoplastic fluids exhibit nearly linear shear stress–shear rate behavior at low shear rates. This is called the 'Newtonian regime.'

Dilatant Flow: The shear stress–shear rate plot of this type of a flow begins at the origin but is characterized by equal increments in the shear stress giving less than equal increments in the shear rate Examples are high solids, raw starch suspensions, and some chocolate syrups. This type of flow is only found in liquids that contain a high proportion of insoluble rigid particles in suspension. Dilatant flow is fairly rare in the food industry and extremely rare in finished food products. This type of flow is described as 'dilatant' because it is associated with an increase in volume of the fluid as flow occurs, and it only occurs in high concentration suspensions. Suspension of cornstarch in water. True dilatancy can probably exist in any suspension so long as the concentration is high enough for the material to exist in closely packed form. The property of dilatancy disappears when the suspension is diluted. For example, a 40% cornstarch suspension in water shows no dilatant properties. The densest packing of spheres is about 74% and one of the least-dense packing is about 37%. Hence it is usual to find that the property of dilatancy only appears in suspensions between about 40% and 70% solids concentration. Some fluids that do not dilate when sheared may still exhibit a dilatant type of shear stress–shear rate behavior; that is, equal increments in shear stress give less than equal increments in shear rate. The general term 'shear thickening' applies to these fluids as well as to dilatant fluids.

Rotational Viscometers

There are some empirical viscometers in which a paddle, a cylinder, or bars rotate in a container, usually with large clearances between the rotating member and the wall. The geometry of these viscometers is complex and usually not amenable to rigorous mathematical analysis. These instruments are generally rugged, moderate in cost, and fairly easy to manipulate. They have their place, particularly for quality control purposes in the plant where detailed mathematical analysis is not needed. They are widely used in industry. Examples of this type are some Brookfield Viscometers, and the FMC Consistometer.

- The ***Brabender Viscograph*** is designed specifically to measure the apparent viscosity of starch suspensions and record how the viscosity changes as the temperature of the water–starch slurry is raised past the gelatinization temperature, held at this elevated temperature for a period, and then cooled again.
- The ***FMC Consistometer*** was originally designed by the Food Machinery Corporation to measure the consistency of cream-style corn. It is now distributed by C. W. Brabender Instruments and has been used for routine quality control purposes for catsup, tomato paste, strained baby foods, and other products that have a similar consistency. The product is placed in a stainless steel cup, the paddle is lowered into the cup, the motor is switched on causing the cup to rotate at a single fixed speed of 78 rpm, and the torque on the paddle is read from a scale on top of the instrument. Four paddles with different dimensions are provided with this instrument. The instrument is approximately 38 cm high, 26 cm wide, 31 cm long, and weighs about 16 kg; the four paddle sizes available are 2 in. 1.4 in., 2 in. 1 in., 2 in. 0.75 in. and 1.5 in 1 in.
- The ***Corn Industries Viscometer*** (CIV) was once widely used to measure changes in viscosity of corn syrup and starch pastes, but this instrument is no longer commercially available.
- The ***Brookfield Dial Reading*** Viscometer is an instrument that may be held in the hand or supported on a stand. A synchronous induction type motor gives a series of speeds of rotation that are constant. Various spindles that take the form of cylinders, disks, and T bars are attached to a small chuck. When the spindle is immersed in the liquid and the motor switched on, the viscous drag of the fluid on the spindle is registered as torque on a dial. A Factor Finder scale provided by the manufacturer enables the operator to quickly convert the dial reading into apparent viscosity.

The company can supply a Helipath stand that automatically lowers the Brookfield Viscometer, thus ensuring that the spindle is continuously moving into previously undisturbed material. This accessory is useful when studying fluids that exhibit time effects or that have a tendency to settle. The Brookfield Engineering Laboratories have over 60 years' experience with viscosity measurement and have a good reputation for helping potential customers identify the precise model of instrument and optimum mode of operation for their own applications. Brookfield Dial Reading Viscometers are widely used in the food field. They have the advantages that they are of moderate cost, portable, simple to operate, well adapted to many viscosity problems, give results quickly, can be used to measure viscosity in almost any container ranging from a 200-ml beaker to a 1000-gal tank, can be used on Newtonian and non-Newtonian liquids, can be used to measure time dependency and hysteresis, are not affected by large particles in suspension, and require minimum maintenance. The disadvantages are that there is a limited range of shear rates, the shear rate can only be changed stepwise, the shear rate varies across the fluid, there can be problems in obtaining shear rate and apparent viscosity for non-Newtonian liquids, and the geometry and flow pattern do not lend them to rigorous mathematical analysis.

Paddle Viscometry

This type of viscometer consists of two or more paddles attached to a shaft that is caused to rotate at one or more speeds while the torque resistance is measured. They may resemble a flag or a star-shaped geometry. They are sometimes called 'mixer viscometers' because the paddles stir and mix the product as they rotate. The stirring action makes them particularly useful for food suspensions containing particulates that settle out. This is an empirical instrument; all test conditions must be kept constant to achieve reproducible results. Researchers are now analyzing the action of this class of instrument and extracting useful rheological parameters even though the shear-rate varies from point to point around the paddles. Rao (1975) pointed out that the average shear stress is directly proportional to the torque and the average shear rate is directly proportional to rotational speed. Hence, the slope of the plot of log (torque) versus log (rpm) will give the value of the flow behavior index n of power law fluids. The consistency index K can be found by testing a fluid with known values of K and n under identical test conditions using the equation:

$$\frac{M_x}{M_y} = \frac{K_x}{K_y}$$

Where, M_x is the measured torque for the test fluid and M_y is the measured torque for the fluid whose properties are known, K_y is the consistency index for the known fluid. K_x can then be computed since M_x, M_y and K_y are known. One restriction for this method is that the flow behavior index n of the known fluid should be about the same value as is n for the test fluid.

The ***Rapid Visco-Analyzer*** is a useful example of mixer viscometry. It consists of a molded plastic stirring paddle attached to an electric motor whose shaft rotates at constant speed and the current required to drive it is constantly monitored by a microprocessor. A disposable aluminum sample container is filled with 4 g of flour and 25 ml of water and placed to surround the paddle, and the motor is started. A split copper block at 96°C rapidly heats the mixture through the gelatinization temperature of the starch. The test takes 3 minutes to perform. The power output correlates with the apparent viscosity. This has become a routine screening test to detect sprout damage to wheat quality when wheat is delivered to grain storage sites.

Falling-Ball Viscometers

This type of viscometer operates on the principle of measuring the time for a ball to fall through a liquid under the influence of gravity. The falling ball reaches a limiting velocity when the acceleration due to the force of gravity is exactly compensated for by the friction of the fluid on the ball. Stokes (1819–1903) was one of the first to study the limiting velocity of falling balls. This is a simple type of instrument that is useful for Newtonian fluids but has limited applicability to non-Newtonian fluids. It cannot be used for opaque fluids because the ball cannot be seen. Stokes' law applies when the diameter of the ball is so much smaller than the diameter of the tube through which it is falling that there is no influence of the wall on the rate of fall of the ball.

Oscillation Viscometry

A vibrating surface in contact with a liquid experiences 'surface loading' because the shear waves imparted to the liquid are damped at a rate that is a function of the viscosity of the liquid. The power required to maintain constant amplitude of oscillation is proportional to the viscosity of the fluid. Oscillation viscometers usually take the form of a stainless-steel ball immersed in the fluid and vibrated at high frequency and low amplitude. This type of viscometer has the advantages of high precision, high sensitivity small changes in viscosity, rapid accumulation of data, and the equipment is easy to clean. The disadvantages are that it operates at one shear rate only. The size of the test sample is not critical so long as it exceeds that volume below which reflection from the walls of the container occurs. This distance is usually less than 5 mm.

References

Bingham, E. C. (1922). Fluidity and Plasticity. McGraw-Hill, New York.

Bourne, M.C. (1982). Food Texture and Viscosity: Concept and Measurement. New York: Academic Press. pp 325.

Bourne, M.C. (1990). Basic principles of food texture measurement. In Dough rheology and baked product texture. Springer, Boston, MA. pp. 331–341.

Bourne, M.C. (2002). Food texture and viscosity: concept and measurement. Elsevier.

Chen, L. and Opara, U.L. (2013). Texture measurement approaches in fresh and processed foods—A review. Food Research International, 51(2): 823–835.

Chung, C. Degner, B. and McClements, D.J. (2012). Instrumental mastication assay for texture assessment of semi-solid foods: Combined cyclic squeezing flow and shear viscometry. Food Research International, 49: 161–169.

Jaworska, G. and Bernas, E. (2010). Effects of pre-treatment, freezing and frozen storage on the texture of Boletus edulis (Bull: Fr.) mushrooms. International Journal of Refrigeration, 33: 877–885.

Jones, D., Chinnaswamy, R., Tan, Y. and Hanna, M. (2000). Physicochemical properties of ready-to-eat breakfast cereals. Cereal Chemistry, 45: 164–168.

Kim, E.H.J., Corrigan, V.K., Wilson, A.J., Waters, I.R., Hedderley, D.I. and Morgenstern, M.P. (2012). Fundamental fracture properties associated with sensory hardness of brittle solid foods. Journal of Texture Studies, 43: 49–62.

Rao, M.A. (1975). Measurement of flow properties of food suspension with a mixer. Journal of Texture studies, 6: 533–539.

Varela, P., Salvador, A. and Fiszman, S. (2009). On the assessment of fracture in brittle foods II. Biting or chewing? Food Research International, 42: 1468–1474.

Williamson, R.V. (1929). The flow of pseudoplastic materials. Ind. Eng. Chem., 21: 1108–1111.

17

Properties of Developed Herbal Burfi

R. K. Goyal, S. K. Goyal and P.K. Singh

Abstract

Herbal Burfi was development with various types of milk viz., Buffalo (6% Fat), Cow (4% Fat) and mixed milk (5% Fat) and different levels of sugar i.e. 20, 25, 30 and 35 per cent (weight basis of khoa). Each sample was fortified with 5g satavar and safed musli powder. Other minor ingredients like Almonds, Cardamoms, Cashew nuts, Pistachio nuts, etc were mixed thoroughly and after preparation herbal burfi was allowed to set at room temperature (20-25°C) for two hours. Physico-chemical (acidity, pH, fat, protein, lactose, sugar, ash and total solids), microbial (total plate count, coliform count, yeast and mould count) quality and sensory attributes of the herbal burfi were evaluated. The data of experiments were statistically analyzed to compare significance effects of various factors by applying factorial Randomized Block Design (fRBD) with the help of OPSTAT software and ANOVA.

Introduction

A huge portion of milk produced in India is converted into milk products. Khoa is an indigenous milk product of considerable economic and nutritional importance. It is estimated that about 50% of the total milk produced is converted into traditional milk products. Traditionally, food products have been developed for taste, appearance, value and convenience of the consumers. The development of new products to confer a health benefit is a relatively new trend, and recognizes the growing acceptance of the role of diet in disease prevention and treatment. Among different terms that have been used to describe the many natural products currently being developed for health benefit viz., nutraceuticals or functional foods are the commonly used terms. However, both the terms are often used interchangeably. As per an estimate, the functional foods market is valued at US $ 70 billion and is projected to grow at three times rate (Singh and Srivastava, 2008). This represents 4% of the processed food market. A recent survey conducted by a leading management firm in South Asia revealed that the health and wellness foods market in India is projected to increase from current Rs. 10,150 crore to Rs. 55,000 crore by the year 2020 (TSMG, 2010).

India is bestowed with a wealth of medicinal plants which are among the biggest repositories of herbal wealth in world. About 2000 indigenous plant species have curative properties. The Indian systems of the medicine have identified 1500 medicinal plants of which about 500 species are commonly used in the preparation of herbal drugs (Anon, 2005). About 12.5% of the 4,22,000 plant species documented worldwide are reported to have medicinal values (Rao, 2005).

The Indian systems of the medicine have identified 1500 medicinal plants of which 500 species are commonly used in the preparation of herbal drugs. Today, more than 70 species of medicinal and aromatic plants are commercially cultivated in India. India's share is about 1.0 per cent in the World Trade Market with over 70 billion dollars. Medicinal plants are used for preparation of pharmaceuticals, herbal food, cosmetic products, perfumes, etc. (Sen, 2004). Now-a-days, some important medicinal and aromatic plants are grown by marginal and small farmers (Anon., 2006). Two herbal plants i.e. Satavar (*Asparagus recemosus*) and Safed musli (*Chlorophytum borivilianum*) having medicinal importances were taken in to account to unearth their worthiness as dairy supplement product. So it becomes imperative to have a glimpse of medicinal profiles of these herbal plants. In this context, the following paragraphs are devoted to elaborate the medicinal profiles of these two herbal plants as here under.

Satavar (*Asparagus recemosus*)

Asparagus is one of the important genera of family Lilicaceae represents around 1500 species worldwide and are distributed in temperate and subtropical regions. It represents highly valuable plant species having therapeutical and nutraceutical importance, known as Satavari. Ayurveda considers it as a powerful rasayan drug capable of improving memory power, intelligence, physical strength and maintaining youthfulness. Plant has also ornamental value both for indoor and outdoor decorations. It is found throughout India, in all districts up to 1500 meters elevation. Asparagus roots contain protein (22%), fat (6.2%), carbohydrate (3.2%), vitamin B (0.36%), vitamin C (0.04%) and traces of vitamin A. It contains several alkaloids. Asparagamine A, a novel polycyclic alkaloid possessing antitumor activity is present in it. It contains four antioxytoxic saponinsshatvarin I to IV. Analysis of root tubers at optimum harvesting stage of two years, recorded total free amino acids (0.429%), soluble sugar (45.07%), insoluble sugar (4.79%) and total saponins (1.77 %) (Kurian, 1999).

Tuber roots are demulcent, diuretic, aphrodisiac, tonic, alterative antiseptic, antidysentric, galactogogue and antispasmodic. Its roots are used in leucorrhoea, seminal debility, general debility, agalactia, headache, hysteria, reduces blood pressure, acidity and ulcer patient, etc. (Brahavarchash, 2006). In Ayurveda, it is prescribed as a cooling agent and uterine tonic. Besides quenching thirst, its roots juice helps in cooling down the body from summer heat, curing hyper-activity and peptic ulcer. It contains good amount of mucilage that smoothes inner cavity of stomach. It relieves burning sensation while passing urine and is used in urinary tract infections. It contains anticancer agent asparagine that is useful against leukemia. It also contains active antioxytoxic saponins that have good antispasmodic effect and specific action on uterine musculature. It is a very good relaxant to uterine muscles, especially during pregnancy and is used to prevent abortion and pre-term labour on place of progesterone preparations. Its powder boiled with milk is generally used to prevent abortion. It is a good remedy for vaginal discharges like leucorrhoea, uterine disorders, and excess of bleeding and colic pain. Its preparations with milk help to increase breast milk in lactating women. Its proper use helps in avoiding excess blood loss during periods. It clears out infections and abnormalities of uterine cavity hence; it is used to rectify infertility in women. The roots have an important ingredient for preparations like Satavarigulam, Shatavarighrtam, Sahacaradikulambu, rasnadikasayam, Dhatryadighrtam, etc.

Safed musli (*Chlorophytum borivilianum*)

Safed musli or Chlorophytum borivilianum (Chloros means green and phyton means plant) is commonly known as Dholi musli and locally pronounced by the tribals of Rajasthan, Madhya Pradesh and Gujrat as Koli, Jhirna (Gharwal), Sepheta musli (Bombay) and Khairuwa (North-West Province) (Bordia *et al.*, 1995). Safed musli is being grown naturally in Himalayan tarai, Uttarakhand, Western Uttar Pradesh, Chattisgarh, Southern part of Rajasthan, Western Madhya Pradesh and North Gujrat (Sharma, 2007). The roots of safed musli are white, smooth and 3-5 inches long. A single plant produces 5-7 tuber roots. The roots of safed musli are used in therapeutic preparations against leucorrhoea. It has found common uses due to its aphrodisiac properties and as a tonic for lactating mothers and for women after delivery. Root is also used in commercial preparation of steroidal hormones. Generally, it is used for increase vitality. Besides this, it is also used in many Ayurvedic preparations prescribed for joint pain, diarrhea, diabetes and as a blood purifier. The tubers of safed musli are also used as a substitute for European salad (Mishra, 1994).

From the ancient times, the rural population used roots, shoots, stems, barks, leaves, flowers, etc. for preparation of traditional medicinal formulations to cure several human ailments. With the advancement of medicinal science and development of newer branches of therapy, people diverted herbal to laboratory based synthetic chemical drugs for quick relief. However, about

80% people of the developed countries are intending towards the plant based medicine system due to higher cost and side effect of the chemical drugs. Developments of dairy products supplemented with herbal ingredients are important from nutritional and therapeutic point of view. Now a days, the demand of milk and milk products are increasing day by day. The productions of such herbal products are more economical and profitable in the interest of health care (Chen *et al.*, 2003). The main sources of herbal ingredients are only the edible medicinal plants. Therefore, there is a great need for preparation of some herbal products having more medicinal values.

A little work has been reported on development of medicinal plants based herbal dairy products. The information regarding development of herbal burfi based on satavar and safed musli are scanty. Therefore, present study was conducted to develop satavar and safed musli based herbal burfi.

Materials and Methods

Raw materials satavar (*Asparagus recemosus*) and safed musli (*Chlorophytum borivilianum* Santapau and Fernades) roots were procured from Ashoka Herbal Garden, Haridwar (Uttarakhand). Milk was procured from Dairy Farm, Department of Animal Husbandry and Dairying, R. B. S. College, Bichpuri, Agra (U. P.). Other ingredients, butter paper and paper board boxes were procured from the local market of Agra for the present study. Whole development work of the herbal burfi was done in the laboratory of Dairy and Animal Husbandry department.

Preparation of Herbal Burfi

The khoa was prepared in the laboratory using the methodology as suggested by Banerjee *et al.*,. (1968). The calculated amount of satavar powder, safed musli powder, cashew nut, almonds, cardamoms and pistachio nuts were added with freshly prepared khoa in karahi. The karahi was put up over the gas flame and stirring was done continuously. Required quantity of sugar was added. Potassium sorbate (1g) was added during stirring. The detailed information of scientific technical work plan is given in Table-1. After proper mixing the whole mixture was transferred in aluminum trays and spread up to 1.0 cm thick layer. The mixture was allowed to cool at room temperature for about 60 minutes then after setting burfi was cut into small pieces of size 3x4 cm and wrapped with butter paper. The process chart for preparation herbal additives/ingredients of medicinal plants and preparation of herbal burfi are given in Fig.-A and Fig.-B, respectively.

Table 1 Scientific and Technical work plan for development of herbal burfi

Experimental parameters	levels	Description
Milk source	3	Buffalo milk : 6.0 % fat and 9.5 % SNF Cow milk : 4.0 % fat and 8.5 % SNF Mixed milk : 5.0 % fat and 9.0 % SNF
Method used for preparation of khoa	1	Standard laboratory method as suggested by Banerjee *et al.*, (1968)
Sugar levels	4	20, 25, 30 and 35 % weight of khoa
Medicinal plants used	2	Satawar and Safed musli
Quantity of herbal additives (powder)	1	@ 10 g per kg of burfi (5 g of each medicinal plant)
Other ingredients	-	Cashew nuts, almonds, cardamom and pista
Preservative: Potassium sorbate	1	@ 0.1 % of khoa (w/w)
Packaging mode	1	Paper board boxes (20 x 15 cm)
Storage conditions	1	Low temperature 5±1°C (refrigerator)
Replication	3	Each experiment was replicated for 3 times
Measuring quality parameters		**Sensory** - Colour, flavour, taste, body & texture and overall acceptability (Nelson and Trout, 1982; Ranganna, 2001) **Physico-chemical** – Acidity, pH, protein, fat, Total solids, lactose, sugar and ash. **Microbial** – Total Plate Count (SPC), coliform count and yeasts and mould count after 0, 3, 6, 9 12 and15 days during storage (Chalmer, 1955).
Statistical design	1	factorial RBD (Gomez and Gomez, 1984), STATPAC (OPSTAT) software and ANOVA

In view of above descriptions the salient features of the experimental treatments are given for easily understanding and explicit presentation.

Milk sources : 3 (Buffalo, Cow and Mixed)
Levels of sugar : 4 (@ 20, 25, 30 and 35 % weight of khoa)
Additive levels : 2 (@ 0 and 10g i.e., 5 g each satavar and safed musli)

As such the details about treatment combinations examined are given here under as.

BS_1 : Buffalo milk (6.0% fat and 9.5% SNF) + 20% Sugar + 10g herbal powder.

BS_2 : Buffalo milk (6.0% fat and 9.5% SNF) + 25% Sugar + 10g herbal powder.

BS_3 : Buffalo milk (6.0% fat and 9.5% SNF) + 30% Sugar + 10g herbal powder.

BS_4 : Buffalo milk (6.0% fat and 9.5% SNF) + 35% Sugar + 10g herbal powder.
BC : Buffalo milk (6.0% fat and 9.5% SNF) + 35% sugar without herbal powder (control).
CS_1 : Cow milk (4.0% fat and 8.5% SNF) + 20% Sugar + 10g herbal powder.
CS_2 : Cow milk (4.0% fat and 8.5% SNF) + 25% Sugar + 10g herbal powder.
CS_3 : Cow milk (4.0% fat and 8.5% SNF) + 30% Sugar + 10g herbal powder.
CS_4 : Cow milk (4.0% fat and 8.5% SNF) + 35% Sugar + 10g herbal powder.
CC : Cow milk (4.0% fat and 8.5% SNF) + 35% sugar without herbal powder (control).
MS_1 : Mixed milk (5.0% fat and 9.0% SNF) + 20% Sugar + 10g herbal powder.
MS_2 : Mixed milk (5.0% fat and 9.0% SNF) + 25% Sugar + 10g herbal powder.
MS_3 : Mixed milk (5.0% fat and 9.0% SNF) + 30% Sugar + 10g herbal powder.
MS_4 : Mixed milk (5.0% fat and 9.0% SNF) + 35% Sugar + 10g herbal powder.
MC : Mixed milk (5.0% fat and 9.0% SNF) + 35% sugar without herbal powder (control).

Analysis of herbal burfi

All standard methods were used for evaluation of various important properties i.e. physico-chemical, microbial and sensory analysis of developed herbal burfi samples as following:

Attributes	Methods used
Physico-chemcial analysis	
Total solids (%)	ICAR (1951) and ISI: 1224 (1981) method
Acidity (%)	ISI:1224 (1981)
pH	ISI: 1224 (1981)
Fat content	BIS method (part–II) 1977
Protein content	AOAC (1960)
Lactose content	Knowles and Watkin (1947)
Sucrose content	ISI: 1224 Part-XI (1981)
Ash content	BIS method (1977)
Microbial analysis	
Total Plate Count, Coliform Count, Yeast and Mould Count	Chalmer (1955)
Sensory	
Sensory analysis	Nelson and Trout (1982)
Statistical analysis	factorial RBD (Gomez and Gomez, 1984),
Statistical design	using STATPAC (OPSTAT) software

Development of Satavar and Safed musli Roots Powder (Herbal additives/ingredients)

Procedure for preparation of herbal powder from satavar and safed musli roots is given in Fig.-A.

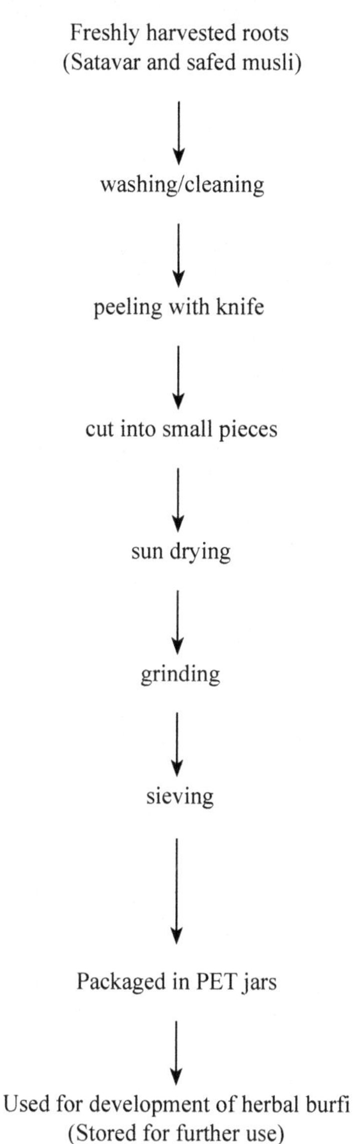

Fig. A Flow diagram for the preparation of herbal/ medicinal additives

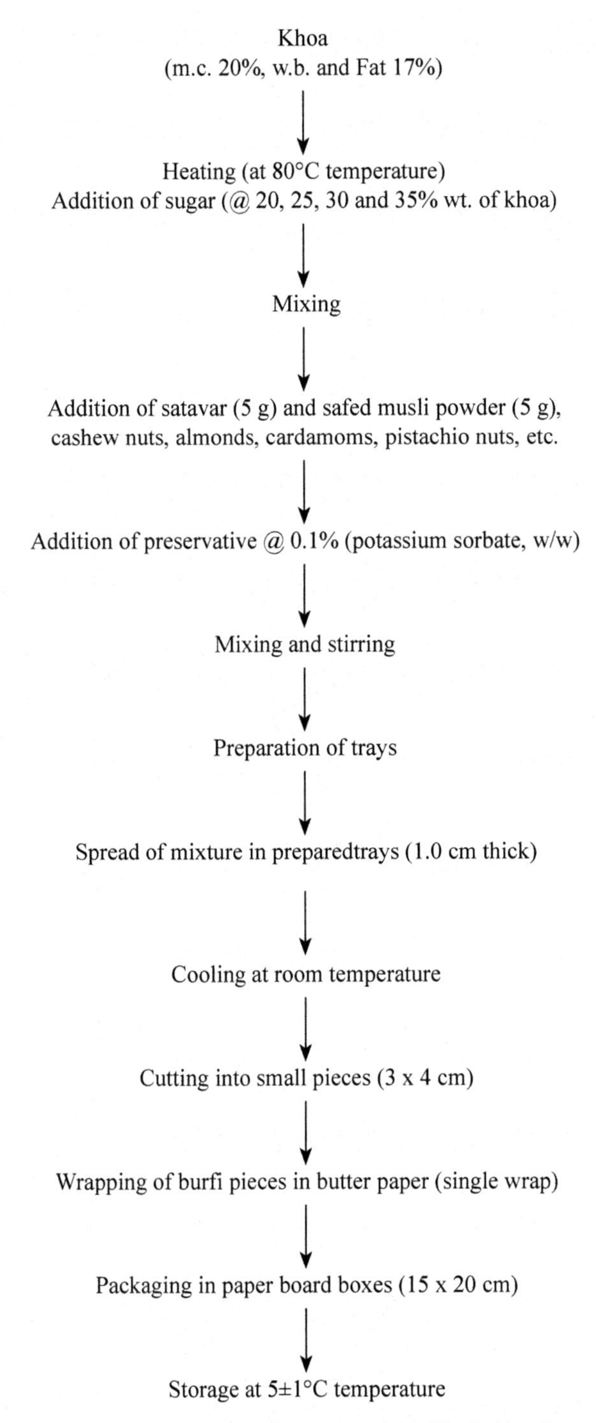

Fig. B: Process chart for preparation of herbal burfi

Results and Discussions

Yield of herbal burfi (%)

The data pertaining to yield parameter of herbal burfi prepared from buffalo milk, cow milk and mixed milk with 20, 25, 30, 35% sugar level and control sample are accessible in Fig. 1. The results clearly indicate that maximum yield of herbal burfi was found in buffalo milk than samples made from mixed milk and cow milk respectively. The maximum yield (30.10%) was registered in the sample of buffalo milk prepared with 35% level of sugar followed by mixed milk (27.40%) and minimum yield was noted in samples prepared from cow milk (25.20%) with same level of sugar. The possible reason for higher yield of herbal burfi in buffalo milk samples might be attributed to higher total solids content in buffalo milk as compared to remaining source of milk which possesses relatively lower total solids content.

Further results also depicted that yield of herbal burfi increased significantly ($P < 0.01$) with increase in sugar levels. However, the yield of herbal burfi in control samples were less than those prepared from 35 per cent sugar level and fortified with herbal powder. The yield of the samples prepared from mixed milk with 20, 25, 30, 35 per cent sugar levels and control were 24.30, 25.70, 26.20, 27.40 and 27.00 per cent respectively. The reason for obtaining high yield in burfi samples prepared with 35 per cent sugar level in all milk sources than control samples might be due to addition of herbal powders and other ingredients.

From statistical analysis of data it is explicit that the main effect of different types of milk as well as sugar levels was highly significant ($P < 0.01$). Whereas, the interaction between types of milk and sugar levels was non-significant.

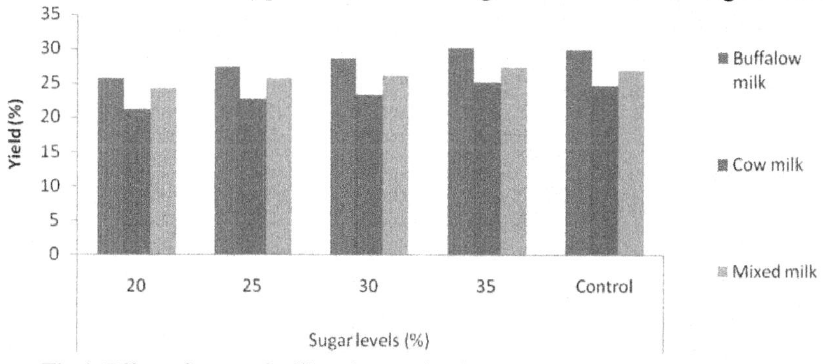

Fig 1 Effect of types of milk and sugar levels on the yield (%) of herbal burfi

Properties of Herbal Burfi

Sensory Property of developed Fresh Herbal Burfi

The results revealed that maximum sensory scores i.e., 8.467, 8.250, 8.000, 8.533 and 7.950 for colour, flavour, taste, body & texture and overall acceptability were found for herbal burfi prepared from buffalo milk sample with 30 per cent sugar which followed by 25, 35 and 20 per cent of sugar levels (Table-2). In case of cow milk the maximum colour scores were noticed in samples prepared by addition of 30 per cent sugar followed in descending order by 25, 35 and 20 per cent sugar levels. Whereas, in case of mixed milk the maximum sensory scores was again encountered with 30 per cent sugar level and followed in descending order by 35, 25 and 20% of sugar levels. The overall acceptability of herbal burfi samples of buffalo milk ranked as maximum (7.950) followed by mixed milk (7.917) and cow milk (7.283), respectively at 30 per cent sugar level in all types of milk. In general, the 30% sugar level obtained maximum sensory score in all three types of milk.

Table 2: Sensory property of developed herbal burfi

Types of milk	Sugar level	Sensory attributes				
		Colour	Flavour	Taste	Body & texture	Overall acceptability
Buffalo milk	BS_1	7.917	6.500	7.000	7.750	7.050
	BS_2	8.050	7.000	7.250	8.250	7.467
	BS_3	8.467	8.250	8.000	8.533	7.950
	BS_4	6.733	7.500	7.750	8.550	7.533
	BC	8.100	8.000	7.433	8.100	7.900
Cow milk	CS_1	6.150	7.300	6.500	7.000	6.800
	CS_2	6.700	6.700	7.000	7.250	6.983
	CS_3	6.900	7.500	7.700	8.000	7.283
	CS_4	6.427	7.000	7.250	7.750	7.143
	CC	6.700	7.500	6.917	7.433	7.017
Mixed milk	MS_1	6.567	6.700	6.883	7.250	7.207
	MS_2	6.667	6.983	7.400	7.500	7.427
	MS_3	6.933	7.280	7.867	8.617	7.917
	MS_4	6.300	7.137	7.583	8.000	7.207
	MC	6.467	6.987	7.433	7.467	7.507
F - ratio		5.28**	4.00**	0.89^{NS}	4.33**	1.12^{NS}
CD (P<0.05)		0.939	0.695	N.S.	0.721	N.S.
Sd		0.456	0.337	0.623	0.350	0.476
±SEm		0.323	0.239	0.440	0.247	0.337
CV (%)		7.975	5.720	10.406	5.475	7.929

* * Significant (P<0.01) NS – non-significant

Physico-chemical properties of developed herbal burfi

The higher values of acidity, protein, fat, lactose and total solids were noticed in buffalo milk burfi samples followed by mixed milk and cow milk at all sugar levels (Table-3). While, maximum sugar (added) was observed in buffalo milk followed by cow milk and mixed milk respectively. In case of pH the highest value was found in mixed milk burfi followed by cow milk and buffalo milk burfi, respectively in 20, 25, 30 and 35% of sugar levels. In general, the values of acidity, protein, fat, lactose, total solids and ash were decreased as sugar level increased in herbal burfi. However, the values of pH and sugar were increased with increase in sugar level.

Table 3 Physico-chemical properties of developed herbal burfi

Sample combinations	Total solids (%)	Acidity (%)	pH	Fat (%)	Protein (%)	Lactose (%)	Sugar (%)	Ash (%)
BS_1	75.417	0.267	6.127	21.167	18.503	21.443	24.060	2.897
BS_2	75.700	0.261	6.180	20.333	17.357	20.637	30.263	2.843
BS_3	75.517	0.256	6.293	19.167	16.380	19.163	35.000	2.770
BS_4	76.860	0.252	6.353	17.833	15.320	18.510	41.120	2.657
BC	77.197	0.250	6.370	18.000	15.023	18.033	41.357	2. 627
CS_1	75.150	0.249	6.190	18.167	13.183	16.813	23.160	3.167
CS_2	75.337	0.247	6.210	17.333	12.540	16.600	29.320	3.143
CS_3	76.000	0.245	6.317	16.167	11.217	15.757	34.600	3.063
CS_4	76.497	0.243	6.353	15.167	11.080	15.400	40.570	2.943
CC	76.983	0.241	6.360	15.333	11.053	15.147	40.867	2.930
MS_1	75.353	0.258	6.160	20.167	16.340	19.103	22.890	2.983
MS_2	75.597	0.254	6.467	19.333	15.623	18.747	29.000	2.903
MS_3	76.453	0.249	6.317	18.167	13.347	18.203	34.270	2.837
MS_4	76.797	0.244	6.393	17.167	12.180	17.997	40.143	2.740
MC	77.113	0.242	6.403	17.500	12.227	17.747	39.300	2.687
F - ratio	12.94**	16.21**	2.39**	88.39**	898.69**	22.13**	634.80**	43.83**
CD (P<0.05)	0.597	0.005	0.191	0.544	0.236	1.117	0.795	0.073
Sd	0.290	0.003	0.093	0.264	0.115	0.542	0.386	0.036
±SEm	0.205	0.002	0.066	0.187	0.081	0.384	0.273	0.025
CV (%)	0.466	1.269	1.803	1.792	0.996	3.700	1.401	1.516

* * Significant ($P < 0.01$)

Microbial property of developed herbal burfi

Results regarding microbial population emerged at fresh condition contained in Fig.-2 clearly illustrate that fresh samples of herbal burfi devoid of coliform and yeast and mould. However, total plate counts had appeared in the herbal burfi samples and the same was enumerated as 2.130, 1.917, 1.817 and 1.773x10^{4}cfu/g in cow milk fortified with 20, 25, 30 and 35% sugar respectively. Whereas, the values for mixed milk herbal burfi samples counted as 1.977, 1.780, 1.607 and 1.600x10^4 cfu/g for respective sugar levels. Minimum number of total plate count were obtained in buffalo milk burfi as the values were 1.783, 1.460, 1.447 and 1.410x10^4 cfu/g. Irrespective of types of milk and sugar levels the total plate count was appeared higher in all three respective control samples with the figures as BC-1.983, CC-2.337 and MC-2.143x10^{4}cfu/g, respectively in comparison with other samples those having different levels of sugar and herbal additives (Fig.-2). In general, total plate count decreased with increase in sugar level in all types of milk. But it further reflects that herbal additives in the form of satavar and safed musli @ 5 g each do appear to have an inhibitory effect on growth of microbial population.

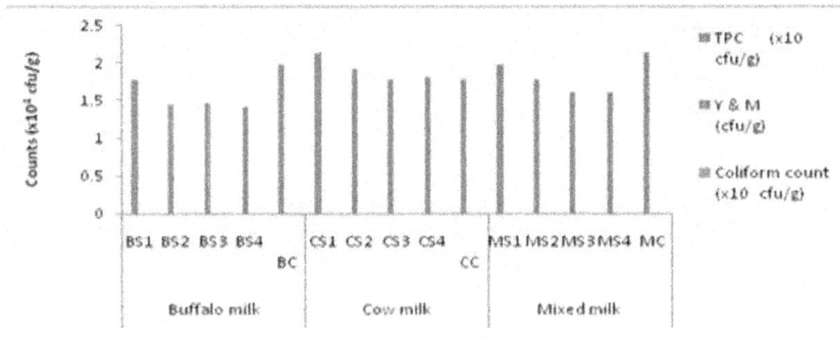

Fig. 2 Microbiological profile of fresh herbal burfi

Cost of Production (per kg) of Herbal Burfi

The comparison and analysis of the different sample combinations of herbal burfi with respect to assessment of the production cost of (Rs./kg) has been obtainable in Fig.-3. It was observed that the lowest cost of production (Rs. 139.20/kg) was enumerated in case of herbal burfi sample made from cow milk with 35% sugar level. However, the highest cost of production (Rs. 169.03/kg) was calculated for samples prepared from buffalo milk with 20% sugar level. The costs of production were decreased significantly ($P<0.01$) with increase in sugar levels for all three types of milk. The cost of production of the samples prepared with buffalo milk using 20, 25, 30 and 35% level of sugar and control were Rs. 169.03, 162.21, 155.38, 148.55 and 132.05 per kg respectively. The possible reason for the lower production cost of samples those having higher sugar level might be because sugar is the cheapest source of solids in comparison to cost of khoa. Therefore, low cost of production (per kg) was found in case of those samples having maximum sugar. In case of control samples the production costs was very low in comparison to samples herbal burfi those containing 35% level of sugar. For example, the production cost (Rs/kg) of buffalo milk, cow milk and mixed milk respective control having 35% sugar was Rs. 148.55 Vs 132.05; 139.20 Vs 122.70; 144.33 Vs 127.67 respectively. The higher production cost of 35% sugar level burfi samples in comparison to control samples of all types of milk may be due to addition of satavar and safed musli powder and preservative in test samples.

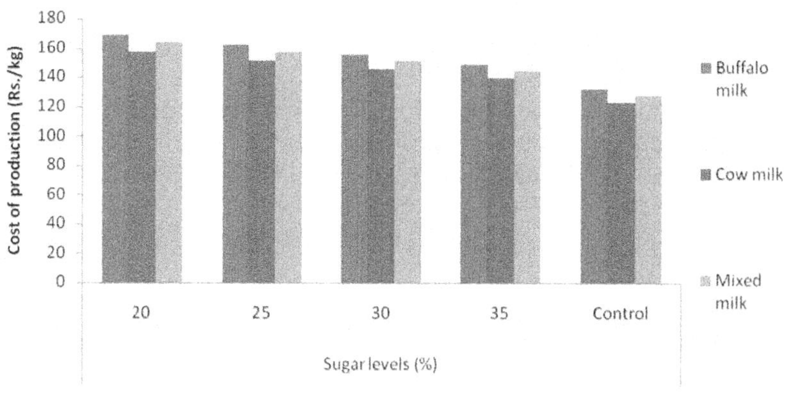

Fig. 3 Effect of types of milk and levels of sugar on cost of production (Rs./kg) of herbal burfii

Conclusion

The maximum yield of herbal burfi was found in buffalo milk followed by burfi made from mixed milk and cow milk. The higher acidity, protein, fat, lactose and total solids were noticed in buffalo milk made herbal burfi followed by mixed milk and cow milk based samples in all sugar levels. While, maximum sugar level (%) were observed in buffalo milk followed by cow milk and mixed milk prepared burfi samples respectively. In case of pH the highest value was found in mixed milk burfi followed by cow milk and buffalow milk, respectively as 20, 25, 30 and 35% of sugar levels. In general, the values of acidity, protein, fat, lactose, total solids and ash were decreased as sugar level increased. However, the values of pH and sugar (%) were increased with increase in sugar level. The coliform, yeast and mould were not detected in fresh samples. The highest total plate count was observed in cow milk burfi samples. Minimum numbers of total plate counts were obtained in buffalo milk. In general, total plate count decreased with increase in sugar level in all types of herbal burfi samples prepared from milk.

In view of the exhaustive and comprehensive results of the present study it could be concluded that buffalo milk having 6.0 per cent fat with 30 per cent sugar appears to be suitable to produce herbal burfi in comparison to other types of milk i.e. cow and mixed milk and other sugar levels. A product with high fat and sugar contents that impart creamy (rich) and sweet taste is preferred by judges. Therefore, the burfi prepared from buffalo milk has superior quality than that made from cow milk. Herbal burfi developed and produced here in present experiments was liked by the judges during its sensory evaluation.

References

Anonymous (2005). Organic farming and medicinal plant cultivation. Agrobios Newsletter, June, pp. 5–6.

Anonymous (2006). Medicinal plants - Export market: Advantage India. The Hindu Survey of Indian Agriculture, pp. 208–211.

Anonymous (2007). Medicinal and aromatic plants: Need for a strategic approach. The Hindu Survey of Indian Agriculture, pp. 186–192.

Anonymous (2011). Pratiyogita Darpan, Indian Economy, Upkar Prakashan, Agra. 105p.

AOAC (1960). Official methods of analysis, Association of Official Analytical Chemist, Washington, DC, 9th Edition, 426 p.

Banerjee, A.K., Verma, I.S. and Bagchi, B. (1968). Pilot plant for continuous manufacture of khoa. Indian Dairyman, 28(2): 81-85.

BIS: 1224 Part – II (1977). Determination of fat by Garber's method – milk products using cheese butyrometer. Bureau of Indian Standards, New Delhi.

Bordia, P.C., Joshi, A. and Simlot, M.M. (1995). Safed musli: Advances in horticulture, Vol.-11 Medicinal and Aromatic Plants. Malhotra Pub. House, New Delhi, pp. 429–451.

Brahavarchash, S. (2006). Ayurveda ka pran : Vanaushadhi vigyan (Hindi). Pub. Yug Nirman yojna, Mathura, pp. 83, 140-148.

Chen, J Lindmark Mansson, H Gorton, L. and Akesson, B. (2003). Antioxidant capacity of bovine milk as assayed by spectrophotometric and amperometric methods. Inter. Dairy Journal. 13(2): 927–935.

Gomez, K.A. and Gomez, A.A. (1984). Statistical procedures for agricultural research. 2nd ed. A Willey International Pub., John Willey & Sons, New York, pp. 8–12.

ICAR. (1951). Standard method for the examination and analysis of milk and milk products other than ghee. Bulletin No. 70: 41–45.

ISI: 1224 (1981). Hand book of food analysis (Part-XI dairy products). Indian Standards Institute Bulletin, Manak Bhavan, New Delhi, 175 p.

Knowles and Watkin. (1947). A practical course in agriculture chemistry, 2nd Ed., 139 p.

Kurian, A. (1999). Final Report of ICAR Adhoc Project "Evaluation and Selection of Medicinal Plants Suitable for a Coconut Based Farming System. Kerala Agri. Univ. Vellankkara, Thrissur.

Mishra, P. (1994). Biochemical investigation of *Chlorophytum* spp. (Safed musli). M.Sc. (Boichemistry) Thesis, (Published), RAU, Udaipur (Rajasthan). 41 p.

Nelson, J.A. and Trout, G.M. (1982). Judging dairy products. Oxford Publisher, U.S.A.

Rao, M.R. (2005). New vistas in agroforestry. Agroforestry system, Pub. Springer Netherlands, pp.107–122.

Sen, N.L. (2004). Importance and scope of medicinal and aromatic plants. In: Manual of winter school and commercial exploitation of underutilized MAPs. MPUA&T, Udaipur, pp. 1–6.

Sharma, A.K. (2007). Medicinal and aromatic plants. Aavishkar Pub. & Distri., Jaipur (Rajasthan). pp. 62–79.

Singh, A.K. and Srivastava, A.K. (2008). Functional foods: From concept to product. In: Lecture compendium of winter school on "Advances in Microbiology, Chemistry and Technology of Dairy Functional Foods and Nutraceuticals", Dairy Microbiology Division, NDRI, Karnal (Haryana), pp: 1–4.

TSMG (2010). Health and Wellness Foods market in India – A report. Accessed on 3rd June, 2010. (http://www.tsmg.com)

18

Physico-Chemical Properties of Edible Oils and Fats

Ratnesh Kumar, Suresh Chandra, Samsher, Vikrant Kumar, Sunil and Vipul Chaudhary

Abstract

Fats and oils are one of the five essential ingredients of human diet and the others are protein, carbohydrates, minerals and vitamins. The oil seeds are major source of edible in world. Oils and fat form an important constituent of human food. Lipid metabolism generates many bioactive lipid molecules, which are fundamental mediators of multiple signaling pathways and they are also indispensable compounds of cell membranes. Many people in developing countries, especially children under five years of age suffer from acute or chronic protein and energy shortages. There is definitely a need for food production to keep pace with the increase in the number of the world's population. Oilseed and nut should be properly dried before storage, and cleaned to remove sand, dust, leaves and other contaminants. All raw materials should be sorted to remove stones and moudly nuts. Some moulds, especially in the case of groundnuts, can cause aflatoxin poisoning.

Introduction

Oil has been a vital part of people's regular dietary consumption all over the world and its usage has been found to increase several folds over the decades. The importance of using the appropriate oil for cooking goes a long way in affecting the consumer's health. Improper methods of oil-aided cooking can lead to cardiovascular diseases and increased cholesterol in the blood (Kumar *et al.*, 2018). There is definitely a need for food production to keep pace with the increase in the number of the world's population. In order to achieve these national development strategies in many based economy tropical countries on agriculture and biased now towards increasing diversity of food products and consumer in order to alleviate malnutrition and pressure to strengthen and expand based industries on agriculture to ensure that their products are both healthy and safe. Oils and fats for food have used a variety of other applications since prehistoric times, and were easily isolate the source of the fats and oils found useful because of its unique properties. These are found ingredients to add flavor, lubricity, and texture, and to satiety foods. It has also been found to have a significant role in human nutrition. Fats and oils is the energy source of the highest of the three basic foods (carbohydrates, proteins, and lipids) and carriers for vitamins soluble oil, and many contain essential for health fatty acids that are not manufactured by the human body (Nelson and Cox, 2005). Vegetable oil is derived from seeds of plants. Among the oilseeds cultivated in India, from which edible oil is obtained, are groundnut, rapeseed, mustard, safflower, sunflower, soybean, linseed. The other sources of vegetable oil are palm, cottonseed, coconut and rice bran. They contain essential fatty acids which play an important role in nutrition and also carriers of fat soluble vitamins. It is estimated that about 90% of vegetable oils are used for edible purposes, while the remaining part finds industrial applications (Kumar *et al.,* 2018). Global oilseed production is projected to rise by 4% during 2013/14. By the year 2020, the production is expected to expand further by 23%, and will be led by United States of America (US), Brazil, China, India, Argentina and European Union (EU). While the biodiesel industry is projected to represent a significant source of demand in the EU and US, in China and India growth occurs primarily in food use. Edible oil consumption in India has nearly doubled to reach 13.4 kg (2010–2011) per capita in the last 5 yrs, making India the second largest consumer (after China); the value is expected to grow further to 24 kg by 2020 (Swati *et al.,* 2015). The quality of any oil is indicated by some physico-chemical properties. The specific value of some of these properties provides an indication of both the nutritive and physical quality of the oil. These properties include iodine value, peroxide value, saponification value, unsaponifiable value, free fatty acid, color appearance etc. For example, oils with low melting point may readily go rancid due to the high level of unsaturation. Recently; palm oil has become the second most consumed oil all over the world as a result of its being rich in natural antioxidants, vitamins and exhibiting high oxidative stability with attendant long shelf life. However, its high melting point lowers the level of acceptance among some consumers (Edem, 2002). The exposure of oils to either a source of heat, light or moisture can alter some of the quality indicators. The extent of alteration (spoilage) depends on the duration of exposure, temperature and condition of storage (Aidos, *et al.,* 2002; Fekarurhobo, *et al.,* 2009).The stability of oil to oxidation is an important indicator in determining oil quality and shelf-life (Choe and Min, 2006) Two common practices that render vegetable oils in most commercial centres prone to oxidative deterioration are packaging in transparent containers (plastic bottles and sachets) and displaying oils either under direct sunlight or artificial lights. Studies have shown that some vegetable oils available in the market do meet the recommended standards for edible oils (Reyes Hernandez, *et al.*, 2007).

Edible / Cooking oil

Cooking oil is plant, animal, or synthetic fat used in frying, baking, and other types of cooking. It is also used in food preparation and flavouring not involving heat, such as salad dressings and bread dips, and in this sense might be more accurately termed edible oil. Cooking oil is typically a liquid at room temperature, although some oils that contain saturated fat, such as coconut oil, palm oil and palm kernel oil are solid. The type of oil will determine the degree of unsaturation or polyunsaturation and the normal level of natural antioxidants, and these factors should be borne in mind when deciding on the requirements for storage. Most other factors affecting degradation are more or less controllable (Jalgaonkar *et al.*, 2017):

- Minimize contact with or absorption of air.
- Avoid excessive moisture.
- Avoid contact with and minimize content of pro-oxidants.
- Keep oils in the dark.
- Keep time and temperature to a minimum.
- Avoid unnecessary agitation.

Properties of edible oils

Colour: The colour of oils by comparison with Lovibond glasses of known colour characteristics. The colour is expressed as the sum total of the yellow and red slides used to match the colour of the oil in a cell of the specified size in the Lovibond Tintometer (Nagaraj, 2009).

$$\text{Colour reading} = Y + 5R \text{ or } Y + 10R$$

Where,

$Y + 5R$ is the mode of expressing the colour of light coloured oils and
$Y + 10R$ is for the dark-coloured oils
Y = Sum total of yellow slides used.
R = Sum total of red slides used.

Specific Gravity / Density: Specific gravity of any substance is it's weight of a unit volume, in comparison to that of water. The weight of 1ml of water is taken as 1g (Nagaraj, 2009). The temperature at which the specific gravity is determined should be specified. Or Specific gravity of oil is determined as the ratio of the density of oil in to the density of water at same temperature.

$$\text{Specific gravity} = \frac{\text{Density of oil}}{\text{Density of water}}$$

Density: The density of oil was calculated by mass of the sample per unit volume.

$$\text{Density} = \frac{\text{mass of the oil(g)}}{\text{volume of the oil (cc)}}$$

Viscosity: The Redwood viscometer is conveniently used for obtaining the relative viscosities of oils, and the time of efflux of a definite volume (usually 50 ml) is determined at the required temperature. In order that data obtained may be comparable, determinations must be made at agreed standard temperatures, usually at 20°C and 60°C. The determination of viscosity is chiefly of importance in the case of oils used for lubricating purposes. The most informative results are yielded in the case of castor, rape and Tung oils (Bolton, 1999).

Viscosity is one of the most important properties of lubricating oil. The formation of fluid film of lubricant between the friction surfaces and the generation of frictional heat under particular condition of load bearing speed and lubricant supply mostly depend upon the viscosity of lubricant and to some extent on its oilness. The viscosity decrease with the rise in temperature as usual therefore variation of viscosity with respect to temperature found in vegetable oils and their blends. This is due to rise in temperature enhances movements of molecule and reduces intermolecular forces so the layer of liquid easily pass over one another and thus contribute to reduction in viscosity. This phenomenon also verified by other researcher since oil viscosity depends on molecular structure and decreases with unsaturation of fatty acid (Talkit *et al.*, 2012).

Smoke, Flash and Fire Points: The smoke point is the temperature at which a fat or oil gives off a thin bluish smoke. It is measured by a standard method in an open dish specified by the American Society for Testing Materials so that the evolution of smoke can be readily seen and reproduced. The flash point is the temperature at which the mixtures of vapor with air will ignite; the fire point is the temperature, at which the substance will sustain continued combustion, for a given sample of oil or fat, the temperature is progressively higher for the smoke point, flash point and fire point. Few oils with flash points measured in an open cup. The temperatures vary with the amount of free fatty acids present in an oil or fat, decreasing with increased free fatty acids. Since the amount of free fatty acids changes with variations in refining, the history of the oil or fat is important. The smoke point of a fat used for deep fat frying decreases with use of the fat. Fats and oils with low molecular weight fatty acids have low smoke, flash and fire points. The number of double bonds present has little effect on the temperature required. Smoke, flash and fire points are particularly useful in connection with fats used for any kind of frying (Meyer, 2000).

Melting Point: The fat is melted and introduced into a tube about 0.025 inches bore, with walls not more than 0.003 inches thick. The fat in the tube must then be thoroughly cooled, either by being kept on ice for half an hour, or allowed to stand at room temperature for at least twenty-four hours. Certain fats, for example, hydrogenated fats, must be allowed to stand for considerably longer periods. Many fats may, however, be introduced into the tube by simply pressing it into the fatty mass, but in certain cases there is a danger of pressing out some of the softer constituents. If the latter method be resorted to and the fat has been solid for twenty-four hours previously, it is obvious that no subsequent cooling is necessary.

The tube is attached to a thermometer by means of rubber bands and the thermometer clamped in position in a beaker containing not less than 500 ml. of water, the top of the tube being submerged. The temperature of the water is now raised by means of a small flame, at the rate of about 0.5oC per minute, the water being kept stirred by mechanical means, preferably by blowing a stream of air through the water. The fat is observed by means of a lens and as soon as it is sufficiently melted to form a clearly defined meniscus on the top, the temperature is read and recorded as the point of incipient fusion. The heating is then continued until the fat is quite free from unmelted particles, this temperature being recorded as the point of complete fusion, which is usually returned as the melting point (Bolton, 1999).

Iodine value: Iodine value is as the grams (number) of iodine absorbed by 100 gram of fat or oil. Iodine number is useful to know the relative unsaturation of fats and is directly proportional to the content of unsaturated fatty acids. Thus lower is the iodine number, less is the degree of unsaturation. The determination of iodine number will help to know the degree of adulteration of given oil (Chandra *et al.*, 2014).

$$\text{Iodine Value} = \frac{(\text{Blank titre} - \text{Sample titre}) \times \text{N of } Na_2S_2O_3}{\text{Weigtht of sample(g)}} \times 12.69$$

Where,

$Na_2S_2O_3$ - Sodium thiosulfate

The iodine value increase at elevated temperature could be attributed to destruction of double bonds in the oils upon heating. This could be due to the effect of high temperature causing destruction of π-bonds and hence decreasing the degree of unsaturation (Ngassapa *et al.*, 2012). These low iodine values may have contributed to its greater oxidative storage stability. The oxidative and chemical changes in oils during storage are characterized by an increase in free fatty acid contents and a decrease in the total unsaturation of oils (Perkin, 1992). The greater the iodine value, the more the unsaturation and the higher the susceptibility to oxidation (Anyasor *et al.,* 2009).

Refractive Index: The ratio of velocity of light in vacuum to the velocity of light in the oil or fat; more generally, it expresses the ratio between the sine of angle of incidence to the sine of angle of refraction when a ray of light of known wave length (usually 589.3 nm, the mean of D lines of Sodium) passes from air into the oil or fat. Refractive index varies with temperature and wavelength. Refractive index generally varies with temperature and wave length of light. Refractive index of oil sample is measured with the help of a suitable refractometer. It increases with the increase in unsaturation and also chain length of fatty acid in oils. It is useful in understanding the structure and purity of oils/fats (Nagaraj, 2009). Refractive Index was determined using a mathematical expression derived by Perkins (1995).

$$RI = 1.45765 + 0.0001164\ IV$$

Where,

RI = Refractive Index, and IV = Iodine Value

The recorded refractive index of the any edible oil under different storage condition with varying temperature were not significantly varied. The values of refractive index obtained for entire sample are same. The higher values of refractive index obtained for the crude oil revealed the necessity to purify the oils. The high refractive index of oil also showed that the fatty acids in oil will contain a high number of carbon atoms (Bello and Olaware, 2012). It is generalized that the refractive index of oils increases with increase the number of double bonds. With increase in temperature, the refractive index of oil decrease. The refractive indices can also be influenced by oxidative damage of oil (Pandurangan, *et al.,* 2014). A slight increase in refractive index of oils during storage may be due to decomposition of saturated fatty acid and compounds which could affect this property.

Peroxide value: The peroxide value is defined as the amount of peroxide oxygen per kilogram of fat or oil. Detection of Peroxide gives the initial evidence of rancidity in unsaturated fats and oils. The double bonds bund in fats and oils play a role in autoxidation. Oils with a high degree of unsaturation are most susceptible to autoxidation. The best test for autoxidation (oxidative rancidity) is determination of the peroxide value. Peroxides are intermediates in the autoxidation reaction. Autoxidation is a free radical reaction involving oxygen that leads to deterioration of fats and oils which form off-flavours and off-odours.

Peroxide value, concentration of peroxide in an oil or fat, is useful for assessing the extent to which spoilage has advanced (Chandra and Kumari, 2016).

Peroxide value = V/W (ml of 0.002N. Sodium thiosulphate per gm)

or

Peroxide value = $2V/W$ (milli equivalent of oxygen per kg)

Where,

V = Volume of 0.002N. $Na_2S_2O_3$ used (ml)
W = Weight of the sample taken (g)

The peroxide value was also found to increase with the storage time, temperature and contact with air of the oil samples. Oils exposed to both atmospheric oxygen and light showed a much larger increase in peroxide value during storage. Increase of peroxide value with storage time has also been reported by Kamau and Nanua (2008). Peroxide value (PV) is used as a measure of the extent to which rancidity reactions have occurred during storage it could be used as an indication of the quality and stability of fats and oils (Ekwu and Nwagu, 2004).

Free Fatty Acid (Acid Value): The acid value of an oil/fat is the number of potassium hydroxide required to neutralize the free acids resulting from the complete hydrolysis of 1g of the sample (Ayo and Agu, 2012).

$$\text{Acid value as oleic acid} = \frac{\text{ml of alkali} \times \text{N of alkali} \times 56.1}{\text{wt of sample(g)}}$$

When the temperature of storage condition increased, the carboxylic acid group may react to from esters and peroxides which degrade to aldehydes, ketones, and other secondary compounds. Hence, the number of available free hydrogen ions reduces (Chen *et al.*, 2013). Higher fat acidity in vegetable oils used in present study might be due to the more content of polyunsaturated fatty acids thereby resulting in breakdown of triglycerides, increasing the free fatty acids which further increases the fat acidity (Preeti *et al.*, 2007). Free fatty acids can stimulate oxidative deterioration of oils by enzymatic and or chemical oxidation to form off flavor components. Free fatty acid value is an indication of lipase activity (Ukhun, 1986).

Reichert-Meissl Value: The Reichert-Meissl (RM) value is the number of millilitres of 0.1N aqueous sodium hydroxide solution required to neutralise steam volatile water soluble fatty acids distilled from 5g of an oil/fat under the prescribed conditions. It is a measure of water soluble steam volatile fatty acids chiefly butyric and caproic acids present in oil or fat.

$$\text{R.M.} = \frac{(A - B) \times 1.10 \times 5}{\text{Weight of the sample(g)}}$$

Where,

A = Volume of N/10 NaOH used in the sample (ml)
B = Volume of N/10 NaOH used in the blank (ml)

Polenske Value: The Polenske value is an indicator of how much volatile fatty acid can be extracted from fat through saponification. It is equal to the number of milliliters of 0.1 normal alkali solution necessary for the neutralization of the water-insoluble volatile fatty acids distilled and filtered from 5 grams of a given saponified fat. It is measure of the steam volatile and water insoluble fatty acids, chiefly caprylic, capric and lauric acids present in oil and fat.

$$\text{Polenske value} = \frac{(C - D) \times 5}{\text{Weight of the sample}}$$

Where,

C = Volume of 0.1 N NaOH used in the sample (ml)
D = Volume of 0.1 N NaOH used in the blank (ml)

Kirschner Value: The Kirschner value is an indicator of how much volatile fatty acid can be extracted from fat through saponification. It consists of the number of milliliters of 0.1 normal sodium hydroxide necessary for the neutralization of water-soluble silver salts made from the water-soluble volatile fatty acids distilled from 5 grams of a given fat.

$$\text{Kirschner Value} = (E - F) \times \frac{(100 + A) \times 121}{10000}$$

Where,

E = Volume of 0.1 N $Ba(OH)_2$ used in the sample (ml)
F = Volume of 0.1 N $Ba(OH)_2$ used in the blank (ml)
A = Volume of 0.1 N NaOH used in the sample determination (When estimating of the R.M. value) (ml)

Saponification Value: Saponification value of an oil or fat is the number of potassium hydroxide required to neutralize the fatty acid resulting from the complete hydrolysis of 1 g of the sample (Ayo and Agu, 2012). The saponification value is the number of milligrams of potassium hydroxide required to sponify of 1g of oil/fat. The saponification value is an index of mean molecular weight of the fatty acids of the glycerides. Lower saponification value indicates higher molecular weight of fatty acid and vice-versa. The oil sample is saponified by refluxing with a know excess of ethanolic KOH. The alkali required for saponification is determined by titration of the excess potassium hydroxide with standard hydrochloric acid (Nagaraj, 2009).

$$\text{Saponification Value} = \frac{(B - S) \times N \times 56.1}{W}$$

Where,

B = ml of N/2 HCl used in the blank
S = ml of N/2 HCl used in the blank
S = ml of N/2 HCl used in the sample
N = Normality of the acid
W = Weight of the sample (g)

The high saponification values indicate that the oils are normal triglycerides and will be useful in the production of soap (Yousefi *et al.*, 2013). Saponification is only of interest if the oil is for industrial purposes, as it has no nutritional significance. But due to the fact that each fat has within the limits of biological variation, a constant fatty acid composition, determination of the saponification value is a reasonable means of characterizing the fat (Tan *el al.*, 2002).

Unsaponification matter: Unsaponification matter is the fraction of substance in oils and fats which is not saponified by caustic alkali, but is soluble in ordinary fat solvents. It includes sterols, higher aliphatic alcohols, pigments, vitamins and hyhrocarbons (Nagaraj, 2009).

Weight (g) of the free fatty acids in the extract as oleic acid = 0.282 VN

Where,

V = Volume of standard sodium hydroxide solution (ml)
N = Normality of standard sodium hydroxide solution

$$\text{Unsaponifiable matter} = \frac{(A - B) \times 100}{W}$$

Where,

A = Weight of the residue (g)
B = Weight of the free fatty acids in the extract (g)
W = Weight of the sample (g)

Acetyl value: Acetyl value is the number of milligrammes of potassium hydroxide to neutralise the acetic acid obtained when 1gem of an acetylated fat or oil is saponified (Bolton, 1999).

$$\text{Acetyl value} = \frac{\text{Number of ml. N / 10 KOH} \times 5.61}{\text{Weight of acetylated fat taken}}$$

Conclusion

Oils and fats for food have used a variety of other applications since prehistoric times, and were easily isolate the source of the fats and oils found useful because of its unique properties. These are found ingredients to add flavor, lubricity, and texture, and to satiety foods. It has also been found to have a significant role in human nutrition. Lipids and triacylglycerol naturally occur in oils and fats. Their chemical composition contains saturated and unsaturated fatty acids and glycerides. Edible oils are vital constituents of our daily diet, which provide energy, essential fatty acids and serve as a carrier of fat soluble vitamins. Cooking oil is typically a liquid at room temperature, although some oils that contain saturated fat, such as coconut oil, palm oil and palm kernel oil are solid. Fat frying is one of the oldest and popular food preparations. Oil has been a vital part of people's regular dietary consumption all over the world and its usage has been found to increase several folds over the decades. The importance of using the appropriate oil for cooking goes a long way in affecting the consumer's health.

References

Aidos, I., Lourenclo, S., Padt, A., Luten, J.B, and Boom, R.M. (2002). Stability of Crude Herring Oil Produced from Fresh Byproducts: Influence of Temperature during Storage. J. Food Sci. 67:3314–3320.

Anyasor, N., Ogunwenmo, O., Oyelana, A., Ajayi, D. and Dangana, J. (2009). Chemical Analysis of Groundnut Oil. Pakistan Journal of Nutrition, 8(3): 269–272.

Ayo, J. and Agu, H.O. (2012). Simplified manual of food analysis for tertiary and research institution, first edition, vol.1, publishing Amana printing and advertising Ltd, 29.

Bellow, M.O. and Olaore, N.O. (2012). Advance food energy security, 4: 632.

Bolton, E.R.R. (1999). Oils, Fats and Fatty Foods. Biotech Books, Teri Nagar, Delhi.

Chandra, S. and Kumari, D. (2016). Short notes on food safety officer examination. Jain brothers, New Delhi.

Chandra, S., Goyal, S.K., Kumari, D. and Samsher (2014). Food Processing andTtechnology (An question bank). Jain brothers, New Delhi.

Chen, W.A., Chiu, C.P., Cheng, W.C., Hsu, C.K. and Kuo, M.I. (2013). Total Polar Compounds and Acid Values of Repeatedly Used Frying Oils Measured by Standard and Rapid Methods. Journal of Food and Drug Analysis, 21(1): 58–65.

Choe, E. and Min, D.B., (2006). Mechanisms and factors for edible oil oxidation. Comp. Rev. Food Sci. and Food Safety, 5:169–186.

Edem, D.O. (2002). Palm Oil: Biochemical, physiological, nutritional, hematological and toxicological aspects: A review. Plant Foods for Human Nutrition (Formerly Qualitas Plantarum) 57:319–341.

Ekwu, F.C. and Nwagu, A. (2004). Effect of processing on the quality of cashew nut oils. J. Sci. Agric. Food Tech. Environ, 4: 105–110.

Fekarurhobo, G.K., Obomanu, F.G. and Maduelosi, J.N. (2009). Effects of Short-term Exposure to sunlight on the quality of some Edible vegetable oils. Research Journal of Applied Sciences. 4 (5):152–156.

Jalgaonkar, K., Sharma, K., Ratnakar, B.B. and Vishwakarma, R.K. (2017). Extraction, Refining and storage of Edible Oils: Methods and Future Expecctations.

Compendium of ICAR-Summer School (Aug 1-21, 2017) on Advanced strategic processing techniques for oilseeds to combat protein-energy malnutrition and augment farmers income. (Ed: Yadav, D.N., Nanda, S.K. and Solanki, C.) ICAR-CIPHET, Ludhiana. pp. 44–56.

Kamau, J.M. and Nanua, J.N. (2008). Agric. Trop. Subtrop. 41(3):106–109.

Kumar, R., Chandra, S., Samsher., Kumar, K., Kumar, T. and Kumar, V. (2018). Analysis of the physical and chemical characteristics of edible vegetable blended oil. International Journal of Chemical Studies, 6(6): 10–15.

Kumar, R., Chandra, S., Samsher., Kumar, V., Kumar, K. and Sunil. (2018). Processing, storability, Physico-chemical properties and human health benefit of edible oil: A review. South Asian Journal of Food Technology and Environment, 4(2): 668–679.

Meyer, L.H. (2000). Food Chemistry. CBS Publishers and Distributors, New Delhi (India).

Nagaraj, G. (2009). Oil Seed Properties, Processing, Products and Procedures. New India Publishing Agency, New Delhi.

Nelson, D.L. and Cox, M.M. (2005). Lehninger Principles of Biochemistry, Fourth Edition, Worth Publishers, Inc. New York, pp: 344.

Ngassapa, F.N., Nyandoro, S.S. and Mwaisaka, T.R. (2012). Effects of temperature on the physicochemical properties of traditionally processed vegetable oils and their blends. Tanz. J. Sci., 38(3): 166–176.

Pandurangan, M.K., Murugesan, S. and Gajivaradhan, P. (2014). Physico-chemical properties of groundnut oil and their blends with other vegetable oils. Journal of Chemical and Pharmaceutical Research, 6(8): 60–66.

Perkin, E.G. (1992). Effect of lipid oxidation on oil and food quality in deep frying. In: Angels, A.J.S. (Ed.), Lipid Oxidation in Food, Chapter 18, ACS Symposium Series no. 500 ACS, American Chemical Society, Washington DC, pp. 310–321.

Perkins, G. (1995). Physical properties of soybeans, in Erickson (Ed.), Practical handbook of soybean processing and utilization, 3 (Champaign, IL: AOCS Press) 29–38.

Preeti., Khetarpaul, N., Jood, S. and Goyal, R. (2007). Fatty acid composition and physico-chemical characteristics of cooking oils and their blends. J. Dairying, Foods & H.S., 26 (3/4) : 202–208.

Reyes- Hernandez, J., Dibildox-Alvarado, E., Charo-Alonso, M. and Toro-Vazquez, J. (2007). Physicochemical and Rheological Properties of Crystallized Blends Containing trans-free and partially Hydrogenated Soyabean Oil. J. Am. Oil Chem. Soc. 84, 1081–1093.

Sawti., Sehwag, S. and Das, M. (2015). A brief overview: Present status on utilization of mustard oil and cake. Indian Journal of Traditional Knowledge,14 (2): 244–250.

Talkit, K.M., Mahajan, D.T. and Masand, V.H. (2012). Study on physicochemical properties of vegetable oils and their blends use as possible ecological lubricant. Journal of Chemical and Pharmaceutical Research, 4(12):5139–5144.

Tan, P., Cheman, B., Jinap, S. and Yousie, A. (2002). Effect of microwave heating on the quality characteristics and thermal properties of RBD palm olein. Innovations in Food Science & Emerging Technology, 3(1): 157–163.

Ukhun, M.E. (1986). Experimental, 42 (8): 94.

Yousefi, M., Nateghi, L. and Rezaee, K. (2013). Investigation of physicochemical properties, fatty acids profile and sterol content in Malaysian coconut and palm oil. Annals of Biological Research, 4(5): 214–219.

19

Chemical and Medicinal Properties of Satavar and Safed Musli

R.K. Goyal, S.K. Goyal, P.K. Singh and Jitendra Kumar

Abstract

The use of plants for treating diseases is as old as the human species. Popular observations on the use and efficacy of medicinal plants significantly contribute to the disclosure of their therapeutic properties, so that they are frequently prescribed, even if their chemical constituents are not always completely known. Medicinal plants have therapeutic potential due to the presence of natural antioxidants functioning as reducing agents, free radical scavengers and quenchers of singlet oxygen. Majority of their antioxidant activity is due to bioactive compounds viz. flavones, isoflavones, flavonoids, anthocyanins, coumarins, lignans, catechins and isocatechins. They have been recognized to possess several medicinal properties (diuretic, expectorant, laxative, anti-bacterial, anti-pyretic etc.) and have been effectively used in the indigenous systems of medicine in India and other countries. Apart from the traditional use, a number of beneficial physiological effects have been identified by extensive studies. Among these are their beneficial effects on lipid metabolism, efficiency as antidiabetics, ability to stimulate digestion and to inhibit platelet aggregation, antioxidant, antilithogenic and anti-inflammatory potential.In this chapter an attempt has made to elucidate and understand the chemical and medicinal properties of Stavar and Safed musli those are using as medicine. These may also use as ingredients for preparation of herbal food products.

Introduction

Medicinal plants are rich in secondary metabolites and are potential sources of drugs. These secondary metabolites include alkaloids, glycosides, coumarins, flavonoids, steroids, etc. Generally the whole plant, roots, stem, bark, leaves, flowers, fruits, gums and oleoresins, etc. are used (Kumar *et al.,* 2006). A large number of people in developing countries have traditionally depended on products derived from plants, especially from forests, for curing human and livestock ailments. Additionally, several aromatic plants are popular for domestic and commercial uses. Collectively they are called medicinal and aromatic plants (MAPs). About 12.5 % of the 4,22,000 plant species documented worldwide are reported to have medicinal values. With dwindling supplies from natural sources and increasing global demand, the medicinal and aromatic plants will need to be cultivated to ensure their regular supply for processing and value added products (Sen, 2004). Recently, the development of food products by supplements with herbal ingredients is important from nutritional and theraptic point of view. Now a day demand of herbal food products is increasing day by day. The production of such type of food product is more economical and profitable in the interest of health (Goyal *et al.,* 2011).

Plants constitute an important source of natural products which differ widely in their structures, biological properties and mechanism of action. Various phytochemical components especially polyphenols, flavonoids, phenolic acids etc. are responsible for the free radical scavenging and antioxidant activity of the plants. Polyphenols possess many biological affects, mainly attributed to their antioxidant activities in scavenging free radicals, inhibition of peroxidation and chelation of transition metals (Nickavar *et al.,* 2006). Over the world range of botanical species and plant parts from which they are derived, they can contribute significant variety and health benefits to the human diet (Goyal *et al.,* 2013).

Importance of medicinal plants

In view of deleterious side effects of synthetic anti-oxidant supplements on human health, the present day's focus is on antioxidants from natural sources (Dapkevicius *et al.,* 1998). In spite of the advent of modern high throughput drug discovery and screening techniques, traditional knowledge systems have given clues to the discovery of valuable drugs (Buenz *et al.,* 2004). Traditional medicinal plants are often cheaper, locally available and easily consumable, raw or as simple medicinal preparations. Considerable research on pharmacognosy, chemistry, pharmacology and clinical therapeutics has been carried out on Ayurvedic medicinal plants in order to establish the scientific basis of their therapeutic potentials (Patwardhan, 2003). Medicinal plants have only recently captured the attention of the scientific community as store house for bioactive compounds providing potential health benefits. As a result, there needs to be significant investment in human clinical trials to substantiate many of the hypothesized health benefits.

Recently, the development of food products by supplementing with herbal ingredients is important from nutritional and therapeutic point of view. The main sources of herbal ingredients are the edible medicinal plants, which are available in rural areas. Some edible medicinal plants like Stevia, Safed musli, Mulhathi, Brhami, Lemongrass, Tulsi, Ginger, Meethineem, Mint, Cardamom, turmeric, Oregano, thyme, marjoram, sage, basil, fenugreek, fennel, coriander, pimento, etc. can be used to develop food products. Various dairy products such as salted spiced buttermilk, whey based herbal beverages, flavored milk, ginger based ice-cream, herbal candies, herbal tea, medicinal ghee, panchamrit, etc are available in Indian market. These products help in curing of physiological disorders, certain diseases and other inborn defects of metabolism in children, young ones and old persons. Herbs and spices are used widely in the food industry as flavours and fragrances. However, they also exhibit useful antimicrobial and antioxidant properties. Many plant derived antimicrobial compounds have a wide spectrum of activity against bacteria, fungi and mycobacteria and this has led to suggestions that they could be used as natural preservatives in foods. Although, more than 1300 plants have been reported as potential sources of antimicrobial agents. India is bestowed with a wealth of medicinal plants which are among the biggest repositories of herbal wealth in world. About 2000 indigenous plant species have curative properties. The Indian systems of the medicine have identified 1500 medicinal plants of which about 500 species are commonly used in the preparation of herbal drugs (Anon, 2005, Rao, 2005).

Need of medicinal plants

The revival of interest in natural drugs, especially those derived from plants, started in the last few decades mainly because of the widespread belief that *"Green medicines"* are healthier and safer than the synthetic ones. Medicinal plants are staging a comeback and an *'herbal renaissance'* is blooming across the world. The WHO took note of the role that traditional medicine which can play in the extension of health services particularly in the remote rural areas (Sen, 2004). Today, more than 70 species of medicinal and aromatic plants are commercially cultivated in India. India's share is about 1.0 per cent in the World Trade Market with over 70 billion dollars (Sen, 2004). According to an estimate, medicinal and aromatic plants occupy an area of about 2,50,000 hectares, producing raw material of worth Rs. 500 crore (Sharma, 2007).

India, with 2.4 per cent of world area and 8.0 per cent of global diversity is one of the 12 mega diversity hot spots. It is believed that one fifth of all plants found in India are medicinally important whereas, the world average stands at about 12.5 per cent (Anon., 2007). In India, medicinal and aromatic plants are cultivated on an area of merely about 0.4 million hectares. Some areas are famous for the cultivation of medicinal and aromatic plants; otherwise most of the produce is collected from forest. Now-a-days, some important medicinal and aromatic plants are grown by marginal and small farmers (Anon., 2006). Two herbal plants i.e., Satavar (*Asparagus recemosus*) and Safed musli (*Chlorophytum borivilianum*) having medicinal importance were taken in to account to unearth their worthiness as dairy supplement product. So it becomes imperative to have a glimpse of medicinal profiles of these herbal plants. It is in this context the following paragraphs are devoted to elaborate the medicinal profiles of these two herbal plants as here under.

Satavari (*Asparagus racemosus F. Liliaceae*)

Asparagus is one of the important genera of family Lilicaceae represents around 1500 species worldwide and are distributed in temperate and subtropical regions. It represents highly valuable plant species having therapeutical and nutraceutical importance, known as Satavari. Ayurveda considers it as a powerful rasayan drug capable of improving memory power, intelligence, physical strength and maintaining youthfulness. Plant has also ornamental value both for indoor and outdoor decorations. It is found throughout India, in all districts up to 1500 meters elevation.

Asparagus roots contain protein (22%), fat (6.2%), carbohydrate (3.2%), vitamin B (0.36%), vitamin C (0.04%) and traces of vitamin A. It contains several alkaloids. Asparagamine A, a novel polycyclic alkaloid possessing antitumor activity is present in it. It contains four antioxytoxic saponins shatvarin I to IV. Analysis of root tubers at optimum harvesting stage of two years, recorded total free amino acids (0.429%), soluble sugar (45.07%), insoluble sugar (4.79%) and total saponins (1.77 %) (Kurian, 1999).

Tuber roots are demulcent, diuretic, aphrodisiac, tonic, alterative antiseptic, antidysentric, galactogogue and antispasmodic. Its roots are used in leucorrhoea, seminal debility, general debility, agalactia, headache, hysteria, reduces blood pressure, acidity and ulcer patient, etc. (Brahavarchash, 2006). In Ayurveda, it is prescribed as a cooling agent and uterine tonic. Besides quenching thirst, its roots juice helps in cooling down the body from summer heat, curing hyper-activity and peptic ulcer. It contains good amount of mucilage that smoothes inner cavity of stomach. It relieves burning sensation while passing urine and is used in urinary tract infections. It contains anticancer agent asparagine that is useful against leukemia. It also contains active antioxytoxic saponins that have good antispasmodic effect and specific action on uterine musculature. It is a very good relaxant to uterine muscles, especially during pregnancy and is used to prevent abortion and pre-term labour on place of progesterone preparations. Its powder boiled with milk is generally used to prevent abortion. It is a good remedy for vaginal discharges like leucorrhoea, uterine disorders, and excess of bleeding and colic pain. Its preparations with milk help to increase breast milk in lactating women. Its proper use helps in avoiding excess blood loss during periods. It clears out infections and abnormalities of uterine cavity hence; it is used to rectify infertility in women. The roots have an important ingredient for preparations like Satavarigulam, Shatavarighrtam, Sahacaradikulambu, rasnadi kasayam, Dhatryadi ghrtam etc.

Kirtikar and Basu (1953) stated that asparagus racemosus wild represents highly valuable plant species having therapeutically and nutraceutical importance in addition to being consumed as vegetable. Known as Satavari, Ayurveda considers it as a powerful rasayana drug capable of improving memory power, intelligence, physical strength and maintaining youthfulness. Plant has also ornamental value both for indoor and outdoor decorations. It is found throughout India, in all districts up to 1500 M elevation. It is also distributed in Africa, through South Asia to China, South Malaysia and Northern Australia (Aiyer and Kolammal, 1962).

Table 1 Variability in Germplasm Accession of Asparagus

S. No.	Characteristics	Range
1.	Plant height (cm)	23.0–382.0
2.	No. of branches	1.0–15.0
3.	No of roots	28.0–355.0
4.	Length of the longest root (cm)	32.0–82.5
5.	Girth of the longest root (cm)	2.75–5.50
6.	Length of the shortest root (cm)	6.0–19.5
7.	Girth of the shortest root (cm)	1.2–3.5
8.	Root weight (g/ plant)	200.0–2650.0

Chemical composition

The major bioactive constituents of *Asparagus* are a group of steroidal saponins. This plant also contains vitamins A, B_1, B_2, C, E, Mg, P, Ca, Fe, and folic acid. Other primary chemical constituents of *Asparagus* are essential oils, asparagine, arginine, tyrosine, flavonoids (kaempferol, quercetin, and rutin), resin, and tannin (Negi *et al.,* 2010). According to Kirtikar and Basu (1975) Asparagus roots contain protein (22%), fat (6.2%), Carbohydrates (3.2%), Vitamin B (0.36%), Vitamin C (0.04%) and traces of vitamin A. It contains several alkaloids. Alcoholic extract yields asparagin, an anticancer agent. Asparagamine A, a novel polycyclic alkaloid passing antitumour activity is also present (Punia *et al.,* 1980; Shah, 1982; Bhattacharya and Ghosal, 1993). It contains four antioxytoxic saponins shatavarin I to IV. Shatavarin IV is a glycoside of sarsapogenin having two molecules of rhamnose and one molecule of glucose. Leaves contain rutin, diosgenin and a flavonoid glycosides identified as quercetin-3-glucuronide (Shivrajan and Balchandran, 1994; Chatterjee and Prakash, 1995). Flowers contain quercetin hypereside and rutin. Fruits contain glycosides of quercetin, rutin and hyperoside while fully ripe fruits contain cyanidin-3-galactoside and cyaniding-3-glucorhamnoside (Warrier *et al.*, 1995). Jain (1996) reported that the analysis of root tubers at optimum harvesting stage of 2 year, recorded total free aminoacids (0.429%),

Soluble sugar (45.0%), insoluble sugar (4.79%) and total saponins (1.77%). Sarasaponin, a steroid drug often used for mitotic chromosome preparation of plant cell is derived from Asparagus species (Chacko, 1997).

Shatavarin IV is a glycoside of sarsasapogenin having 2 molecules of *Asparagus* rhamnose and 1 molecule of glucose (Fig. 1). The major bioactive (Chemical constituents) of *Asparagus* species are shown in Fig. 2). Sarsasapogenin and shatavarin I-IV are present in roots, leaves, and fruits of *Asparagus* species. Synthesis of sarsasapogenin in the callus culture of A. racemosus was also reported by Kar and Sen (1985). A new isoflavone, 8-methoxy-5,6,4'-trihydroxyisoflavone-7-O-β-D-glucopyranoside was also reported from *A. racemosus* previously. The isolation and characterization of polycyclic alkaloid called asparagamine (Saxena and Chaurasia, 2001), a new 9,10-dihydrophenanthrene derivative named racemosol and kaempferol were also isolated from the ethanolic root extract of *A. racemosus (Sekine et al., 1994). Oligofurostanosides* (curillins G and H) and spirostanosides (curilloside G and H), (Fig. 2) have been isolated from the roots and sarsasapogenin from leaves of *A. curillus* (Sekine et al., 1997).

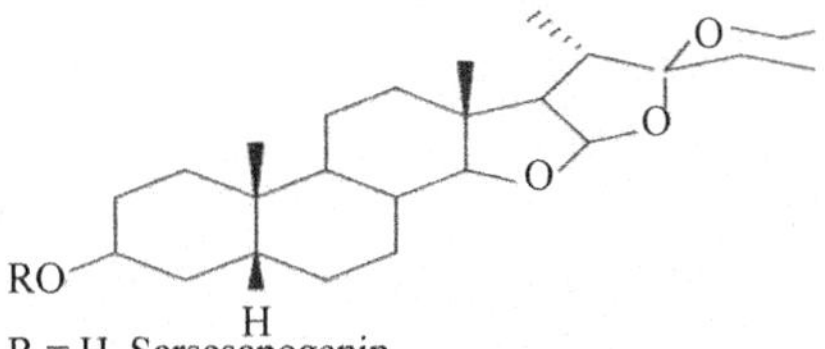

R = H, Sarsasapogenin
R = Glu [(4 - 1) Rha](2-1)Glu, Shatavarin

Fig.1 Structures of sarsasapogenin and its glycosides

R = Glu (6-1), Adscendin A
R = Glu[(4-1)Rha](6-1)Rha, Adscendin B

R = Glu(2-1)Glu, Asparanin A
R = Glu[(2-1)Glu](4-1)Rha, Asparanin B
R = Glu[(4-1)Ara](6-1)Rha, Asparanin C
R = Glu[(2-1)Rha](4-1)Ara, Curillin G
R = Glu[(2-1)Glu](4-1)Ara, Curillin H

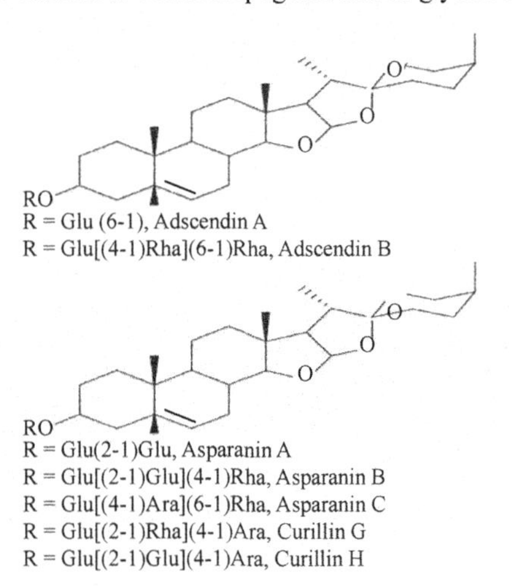

Fig.2 Isolated compounds from Asparagus species

The structural complexity of saponins results in a number of physical, chemical, and biological properties. Saponins are usually amorphous substances having a high molecular weight. These are soluble in water and produce foam but organic solvents, such as chloroform, acetone, and ether inhibit their foaming property. Solubility of saponins is also affected by the properties of the solvent (as affected by temperature, composition, and pH), whereas water, alcohols (methanol, ethanol), and aqueous alcohols are the most common extraction solvents for saponins. Due to the presence of a lipid-soluble aglycone and water soluble sugar chain in their structure (amphiphilic nature), saponins are surface active compounds with detergent, wetting, emulsifying, and foaming properties. In aqueous solutions surfactants form micelles above a critical concentration called critical micelle concentration (cmc).

Saponins possess a variety of biological properties, namely, being antioxidants, immunostimulants, antihepatotoxic, antibacterial, and useful in diabetic retinopathy, anticarcinogenic, antidiarrheal, antiulcerogenic, antioxytocic, and reproductive agents. Many saponins are known to be antimicrobial to inhibit mould and to protect plants from insects. They may be considered as defense system and have been included in a large group of protective molecules found in plants named phytoanticipins or phytoprotectants. Saponin-rich plants have been found to improve growth, feed efficiency, and health in ruminants (Mader and Brumm, 1987)

Medicinal uses

Shrivastava and Pahapalkar (1997) stated that the tuber is demulscent, diuretic, aphrodisiac, tonic, altrative, anticeptic, antidysentric, galactogogue and antispasmodic. In Ayurveda, it is prescribed as a cooling agent and uterinetonic. Besides quenching thirst, its root juice help in cooling down the body from summer heat, curing hyper-acidity and peptic ulcer (CSIR, 1998 and Kurian, 1999).

The roots of satavar contains good amount of mucilage that soothes inner cavity of stomach. It relieves burning sensation while passing urine and is used in urinary tract infections (Farooqui *et al.,* 1999; Rastogi and Mehrotra, (1999). It contains anticancer agent asparagines that are useful against leukemia (Singh *et al.,* 2000; Sanghi and Sharma, 2001). Roots also contains active anti oxytoxic saponins that have good antispasmodic effect and specific action on uterine musculature (Singh and Bangchi, 2002). It is very good relaxant to uterine muscles, especially during pregnancy and is used to prevent abortion and pre-term labour on place of progesterone preparations. Its powder boiled with milk is generally used to prevent abortion (Ates and Erdogrul, 2003). It is a good remedy for vaginal discharges like leucorrhoea, uterine disorders and excess of bleeding and colic pain (Dubey *et al.*, 2004). Its preparation in milk helps in increasing breast milk in lactating women (Ravindran *et al.*, 2004). Its proper use helps in avoiding excessive blood loss during periods. It clears out infections and abnormalities of uterine cavity and hence it is used to rectify infertility in women (Anon., 2005). The blanched young shoots are eaten raw, made into a preserve and candied. The roots form an important ingredient of preparations like Satavrigulam, Satavarighrtam, Sahacaradi Kulambu, Rasnadi Kasayam, Dhatryadighritam, etc. (Brahavarchash, 2006).

Results of research studies showed that Shatarin IV in doses of 20 μg to 500 μg ml^{-1} blocked induced and spontaneous mobility of rabbit uterus. Aqueous extract of roots has inhibitory activity on hatching of eggs of *Meloidogyne arenaria* and *Meloidogyne javanica*. Fresh roots are used to feed the cattle for increase the milk yield (Pandya and Shambhudas, 2007).

Safed musli (*Chlorophytum borivilianum*)

Safed musli or *Chlorophytum borivilianum* (Chloros means green and phyton means plant) is commonly known as Dholi musli and locally pronounced by the tribals of Rajasthan, Madhya Pradesh and Gujrat as Koli, Jhirna (Gharwal), Sepheta musli (Bombay) and Khairuwa (North-West Province) (Bordia *et al.*, 1995). Safed musli is being grown naturally in Himalayan tarai, Uttarakhand, Western Uttar Pradesh, Chattisgarh, Southern part of Rajasthan, Western Madhya Pradesh and North Gujrat. It is distributed in the eastern part of Himalaya, Assam and Bihar. Generally, it grows in certain specified pockets on sloppy hill and widely distributed in tropical and subtropical regions. Commercial cultivation of safed musli has not caught up in our country. The plant grows wild in its natural habitat hence no statistics regarding area and production is available (Sharma, 2007).

The crop of safed musli takes 80-90 days to reach maturity. At maturity, leaves loose green colour, turn yellow and dry plants can be harvested at this stage if the planting material is not required for planting next season. Otherwise, harvesting is done in March-April. Each plant is dug carefully and tubers are washed with clean water. From October-November harvested crop all fleshy tubers are removed from the disc and peeled off and dried in shade. From March-April harvested crop, depending upon planting material requirement, 50-70 per cent of lengthy and thick tubers are removed from the disc and the remaining small and thin tubers (fingers) along with disc are stored for planting in next cropping season. Yield of fleshy roots varies from 5 to 8 tonnes per hectares, which on drying gives 0.6 to 1.2 tonnes. Dried root tubers are stored safely in airtight polypropylene bags with minimum loss of saponin content up to one year (Kurian and Sankar, 2007).

Safed musli roots

The roots of safed musli are white, smooth and 3-5 inches long. A single plant produces 5-7 tuber roots. The roots of safed musli are used in therapeutic preparations against leucorrhoea. It has found common uses due to its aphrodisiac properties and as a tonic for lactating mothers and for women after delivery. Root is also used in commercial preparation of steroidal hormones. Generally, it is used for increasing vitality. Besides this, it is also used in many Ayurvedic preparations prescribed for joint pain, diarrhea and diabetes and also used as a blood purifier. The tubers of safed musli are also used as a substitute for European salad (Mishra, 1994).

A large number of people live in the surrounding of the forest for their subsistence. From the ancient times, the rural population used roots, shoots, stems, barks, leaves, flowers, etc. for preparation of traditional medicinal formulations to cure several human ailments. With the advancement of medicinal science and development of newer branches of therapy, people diverted herbal to laboratory based synthetic chemical drugs for quick relief. However, about 80% people of the developed countries are intending towards the plant based medicine system due to higher cost and side effect of the chemical drugs. The traditional knowledge gained over the period of time passed to the next generations through practices. Due to the nature of plant biodiversity of any region, such knowledge is very much specific to a particular locality.

The name *Chlorophytum* is derived from Greek word, *Chloros* meaning green and *phyton* pant (Rochford and Grover, 1961). It belongs to family Liliaceae. Kirtikar and Basu (1953) had classified it as a medicinal plant. A few *Chlorophytum* species, viz., *C. arundinaceum* Baker (Sharma, 1978; Mishra, 1979; Chadha *et al.*, 1980), *C. attenuatum* Baker (Chadha *et al.*, 1980), *C. borivilianum* Santapau and Fernades (Patel *et al.*, 1991), *C. breviscapum* Dalaz (Singh, 1974), *C. laxum* R. Br. (Mishra, 1979) and *C. tuberosum* Baker (Chadha *et al.*, 1980) are all grouped under one trade name of Safed musli which is used extensively in Ayurvedic medicines.

The genus *Chlorophytum* represents about 175 taxa of rhizomatous herbs distributed in the tropical parts of the world. The probable centre of origin is suggested as the tropical and subtropical Africa where 85 per cent of the species are found. Some of the species are cultivated for ornamental flowers. Thirteen species are reported to occur in India. At least 6-7 *Chlorophytum* species are supposed to be collected and treated as safed musli. The species *Chlorophytum borivilianum* came into prominence during 1980's. Each of the medicinal *Chlorophytum* species reported from India has a specific area of occurrence (Chadha and Gupta, 1995).

Safed musli (*Chlorophytum borivilianum*) is commonly known as Dholi musli and locally pronounced by the tribals of Rajasthan, Madhya Pradesh and Gujrat as Koli, Jhirna (Gharwal), Sepheta musli (Bombay) and Khairuwa (North-west province). The genus *Chlorophytum* (Anthericaceae) with about 215 species is widely distributed in the tropical and subtropical regions. Herbs with short root stock emitting many fascicled roots, often thick, fleshy and tuber like. Leaves are radical, clustered, often broad rarely linear or laureate. White flower is of little horticultural value (Mishra, 1994). The economic part of herb is roots (Bordia *et al.,* 1995).

It is a small perennial herb with radial leaves. The roots tubers are fascicled, sessile, cylindrical, 1-8 in number, brown to black and white after peeling. The tubers are 3-10 cm long at maturity. Leaves are radical, 6-3 in number, spirally imbricate at the base, sessile with acute apex. The leaves are spreading horizontally; margins are waxy with parallel venation. It has a solitary scape, 15-30 cm long, terminal and unbranched (Chadha and Gupta, 1995).

Safed musli is being grown naturally in foot hills of Himalayan, Uttaranchal, Chattisgarh, Southern part of Rajasthan, Western Madhya Pradesh, Western Uttar Pradesh and North Gujarat. It is distributed in the Eastern part of Himalaya, Assam and Bihar (Singh and Chauhan, 2003). Generally it grows in certain specified pockets on sloppy hills and widely distributed in tropical and subtropical regions. At present it is available only in those areas which are not accessible to cattle (Sharma, 2007).

In recent years, many farmers are getting attracted and showing interest in its cultivation on a large scale due to very high profitability. According to an estimate safed musli is cultivated on an area of 20 ha and its cultivation is spreading fast in other areas such as Kota, Jaipur and Jhalawar district of Rajasthan (Bordia, 1991b). Kurian and Sankar (2007) reported that the safed musli *Chlorophytum borivilianum* is an Ayurvedic and Unani medicinal plant since ancient times. The plant is annual herb with sub-erect lanceolate leaves and tuberous roots systems. It is grown for vital natural oil present in the roots, which is used in preparation of health and sexual tonics (Singh and Singh, 2007).

Proximate composition of safed musli

According to Seth *et al.,* (1991), Bordia *et al.,* (1995); Kurian and Sankar (2007) the major constituents of safed musli are carbohydrates (42%), proteins (8- 9%), root fibers (3-4%), saponins (2-17%). Chemically, safed musli constitutes carbohydrates, proteins, fiber and medicinal elements like steroids, saponins, sapogenins and minerals (Singh and Singh, 2007). Mishra (1994) found 4.0 per cent saponin in dry roots of *Chlorophytum borivilianum.* The sapogenin (i.e. only hecogenin) was 0.18 per cent, while the sugar moiety constituted 3.80 per cent. The galactose (0.73%), glucose (0.76%), xylose (0.74%), arabinose (0.79%) and rhamnose (0.78%) were found to be present in saponin hydrolyzate. He also analyzed the safed musli (*Chlorophytum borivilianum*) and elucidated its chemical profiles as delineated hereunder. Safed Musli contains carbohydrates (35- 45%), fiber (25-35%), alkaloids (15-25%), saponins (2-20%), and proteins (5-10%). It is a rich source of over 25 alkaloids, vitamins, proteins, carbohydrates, steroids, saponins, potassium, phenol, resins, mucilage, and polysaccharides and also contains high quantity of simple sugars, mainly sucrose, glucose, fructose, galactose, mannose and xylose (Desale, 2013).

Table-2 Proximate Composition of safed musli

S. No.	Constituents	Value (%)
1.	Saponin	04.4
2.	Total sugar	39.1
3.	Reducing sugar	22.2
4.	Non-reducing sugar	16.9
5.	Total protein	08.5
6.	Water soluble protein	05.8
7.	Crude fiber	05.0

Medicinal Values

Kirtikar and Basu (1953) stated that the dried roots of safed musli are used in therapeutic preparations against leucorrhoea. It has common use due to aphrodisiac properties and as a tonic, in general for lactating mothers and for women after delivery. The possibility of using dietary saponin for reducing liver and serum cholesterol levels has been studied by Newman *et al.,* (1958). The wide ranges of usages of saponins are in soft drinks, beers, confectionary, shampoos, soaps, etc. It have also been used for other pharmaceuticals (George, 1965) like preparations of antihemorrhoidal ointments (Rocher, 1965).

Sambasiva *et al.,* (1982) reported that *Chlorophytum arundinaceum* facilitates sex behaviour in males besides other indigenous plants such as *Mucuna prurita* and *Eulophia campestris.* Tarafder (1983) studied the plants which are used for abortion. The roots of safed musli are white, smooth and 3-5 inches long. A single plant produces 5-7 tuber roots. The roots of safed musli are used in therapeutic preparations against leucorrhoea. It has been reported that due to its aphrodisiac properties it is used as a tonic for lactating mothers and for women after delivery. Root is also used in commercial preparation of steroidal hormones. Generally, it is used for increasing vitality. Besides this, it is also used in many Ayurvedic preparations prescribed for joint pain, diarrhea, diabetes and also used as a blood purifier. The tubers of safed musli are also used as a substitute for European salad (Mishra, 1994).

The dried safed musli root is well known tonic and aphrodisiac drug given to cure general debility. Roots are generally used in the powder form. Tribal's in the central India use leaves of this herb for vegetables purpose. We found no specific work on any of its chemical ingredients which may be responsible for its medicinal property in the literature reviewed (Bordia *et al.,* 1995). *Chlorophytum borivilianum* Baker (Antharicaceae) commonly referred as 'Safed Musli' has been widely used in the Indian traditional systems of medicine to treat various diseases like rheumatism apart from having immunomodulating property and is used as general tonic. It is also known as 'Ayurvedic viagra' for its aphordisiac properties. *C. borivilianum* was screened for the first time to determine its antioxidant activity, isolation of the sapogenins and standardization of the isolated sapogenin fraction using HPLC (Govindarajan *et al.,* 2005).

Kaushik (2005) reported that the genus *Chlorophytum* (Liliaceae) owing to the presence of pharmacologically important saponins has attracted interest of the scientific community to investigate the chemistry of the saponins and study their cytotoxicity. Chloromaloside-A having cytotoxicity against cancer cell lines has been isolated from *C. malayense*, while saponins from *C. borivilianum* are gaining popularity as substitute for viagra. It is a dietary supplement recommended for prenatal and post-natal problems in women. It is known for cure of diabetes, arthritis and gynaecological problems. It also improves the overall immunity of the body. Its consumption is considered an effective means to slow down ageing in humans (Singh and Singh, 2007).

Applications for Male (Desale, 2013)

1. Controls erectile dysfunction due to any reason whether psychological reasons or health problems.
2. Stops and cures premature ejaculation.
3. Increases sexual desire and overcome frustration and embarrassment over a sexual dysfunction, especially erectile problems.
4. Improves energy.
5. Provides ability to maintain the erection throughout the sexual act.
6. Improves physical power and stamina.
7. Improves the semen quality.
8. Increases sperm count substantially (Plays a vital role in Fertility).
9. Increases frequency of orgasm.
10. Solves erectile problems.

11. Increases the volume of ejaculation.
12. Used for faster recovery for second orgasm.
13. Provides extra time, extra pleasure, and extra satisfaction in sexual act.
14. Controls premature ejaculation.
15. Improves and promotes general well being and vitality.
16. Helpful in prolong performance.
17. Increases libido.
18. Equally good for male and female.
19. Increases sexual confidence.
20. Yearlong action with the same intensity.

Applications for Female

1. Controls Erectile dysfunction due to any reason whether psychological reasons or health problems.
2. Increases sexual desire and overcome frustration and embarrassment over a sexual dysfunction.
3. Improves energy.
4. Checks veganism's (tightness of Vagina).
5. Checks for vaginal dryness.
6. Improves physical power and stamina
7. Improves the vaginal lubrication.
8. Increases frequency of desire.
9. Checks the menopause.
10. Maintains youthfulness.
11. Checks urinary tract infections.
12. Checks excessive menstrual bleeding.
13. Checks hormonal shifts and imbalances.
14. Provides full sexual satisfaction throughout the sexual act.
15. Checks lack of willingness.
16. Able to respond to natural sexual urges which is a leading cause of
17. Nervous disorders in women.
18. Improves and promotes general well being and vitality.
19. Develops curiosity towards the sexual act.
20. Increases libido.
21. Checks the loss of sensation in sexual organs.

Conclusion

Since ancient times, plants have been used by several communities to treat a large number of diseases, including infections. Numerous studies on the pharmacology of medicinal plants have been accomplished, since they constitute a potential source for the production of new medicines and may enhance the effects of conventional antimicrobials, which will probably decrease costs and improve the treatment quality. However, several plants may present antagonistic effects during antibiotic therapy. An important aspect comprises the search for new compounds that have antimicrobial action and synergism with currently available antimicrobial drugs, since bacteria resistant to conventional medicines are increasingly frequent; consequently, medicinal plants constitute an alternative for infection treatment. Such medicinal plants may widely used as ingredients for development of new herbal food products.

References

Aiyer, K.N. and Kolammal, M. (1962). Pharmacognosy of Ayurvedic Drugs. Department of Pharmacognosy, Thiruvananthapuram. 65.p.

Anonymous (2005). Organic farming and medicinal plant cultivation. Agrobios Newsletter, June, pp. 5–6.

Anonymous (2006). Medicinal plants - Export market: Advantage India. The Hindu Survey of Indian Agriculture, pp. 208–211.

Anonymous (2007). Medicinal and aromatic plants: Need for a strategic approach. The Hindu Survey of Indian Agriculture, pp. 186–192.

Ates, D.A. and Erdogrul, O.T. (2003). Antimicrobial activities of various medicinal and commercial plants extracts. Turkish J. Biol. 27(3): 157–162.

Bhattacharya, S.K. and Ghosal, S. (1993). Environmental evaluation of the antistress activity of a herbal formulation, Zeestrees. Indian J. Indigenous Medicine, 10(2): 1–8.

Bordia, P.C. (1991b). Propagation studies in Safed musli (Chlorophytum spp.). All India Coordinated Project on Medicinal and Aromatic Plants (ICAR), IX Workshop Report. Gujarat Agril. Uni., Anand 12-15 Dec., pp. 31–42.

Bordia, P.C., Joshi, A. and Simlot, M.M. (1995). Safed musli: Advances in horticulture, Vol.-11 Medicinal and Aromatic Plants. Malhotra Pub. House, New Delhi, pp. 429–451.

Brahavarchash, S. (2006). Ayurveda ka pran : Vanaushadhi vigyan (Hindi). Pub. Yug Nirman yojna, Mathura, pp. 83, 140–148.

Buenz E.J., Schnepple D.J., Bauer B.A., Elkin P.L., Riddle J.M. and Motley T.J. (2004). Techniques: bioprospecting historical herbal texts by hunting for new leads in old times. Tren Pharmacol Sci.; 25: 494–98.

Chacko, K.C. (1997). Flowering, fruiting and seed characteristics of Asparagus racemosus. Journal of Economic and Taxonomic Botany, 21(1): 113–116.

Chadha, K.L. and Gupta, R.C. (1995). Advances in Horticulture, Vol. 11, Medicinal and Aromatic Plants. Malhotra Pub. House, New Dehli.

Chadha, Y.R.; Gupta, R. and Nagarajan, S. (1980). Scientific appraisal of some commercially important medicinal plants of India. Indian Drugs and Pharma. Indus., 15(2): 78.

Chatterjee, A. and Prakash, S.C. (1995). The Treatise on Indian Medicinal Plants. Vol. 1-4. Publication and Information Directorate, CSIR, New Delhi.

CSIR. (1998). The Wealth of India-Raw Materials, Vol. 1-11. Council of Scientific and Industrial Research, New Delhi.

Dapkevicius, A.; Venskutonis, R.; Van Beek, T.A. and Linssen, J. P. H. (1998). Antioxidant activity of extracts obtained by different isolation procedures from some aromatic herbs grown in Lithuania. J Sci Food Agric.; 77:140–46.

Desale, P. (2013). Safed Musli: Herbal Viagra for Male Impotence. Journal of Medicinal Plants Studies, 1(3): 91–97.

Dubey, M.K.; Kumar, R. and Tripathi, P. (2004). Global promotion of herbal medicine-India's opportunity. Curr. Med. 86: 37–41.

Farooqui, A.A. Khan, M.M. and Vasundhara, M. (1999). Productive Technology of Medicinal and Aromatic Crops. Natural Remedies Pvt. Ltd., Bangalore, pp. 55–56.

George, A.J. (1965). Legal status and toxicity of saponin. Food & Cosmet. Toxicol. 3: 85–91.

Govindarajan, R, Sreevidya, N. Vijayakumar, M. Dixit, V.K. Mehrotra, S. and Pushpangadan, P. (2005). In-vitro antioxidant activity of ethanolic extract of chlorophytum borivilianum. India J. Natur. Prod. Sci. 11(3): 165–169.

Goyal, R.K.; Singh, P.K.; Goyal, S.K. and Somvanshi, S.P.S. (2011). Uses of Herbal Ingredients in Food Products. In: National Seminar on Recent Advances and Future Challanges in Ayurvdea held on 26th March, 2011 at BHU, RGCS, Barkachha, Mirzapur.

Goyal, S.K.; Singh, Shree Ram; Singh, S.P.; Rai, J.P.; Singh, S.N. and Prasad, Sant (2013). Antimicrobial and Anioxident properties of Herbs and Spices. In: Global Scenario of Traditional System of Medicine, Ayurveda, Agriculture and education held on 21-22 Jan., 2013 at BHU, RSGC, Barckachha, Mirzapur, pp. 536–538.

Jain, S.K. (1996). Medicinal Plants. National Book Trust, New Delhi, India

Kar, D.K. and Sen, S. (1985). Sarsasapogenin in callus culture of Asparagus racemosus. Curr Sci.; 54:585.

Kaushik, N. (2005). Saponins of Chlorophytum spp. J. Phytoche. (Reviews), pp. 191–196.

Kirtikar, K.R. and Basu, B.D. (1953). Indian Medicinal Plants Vol. lll. Lalit Mohan Basu, Allahabad, pp. 2033–2044.

Kirtikar, K.R. and Basu, B.D. (1975). Indian Medicinal Plants Vol. l to lV. Prakash Publishers, Jaipur.

Kumar, R.; Singh, A. and Thakur, V. S. (2006). Medicinal and aromatic plants of cold desert areas in Himachal Pradesh. Agrobios Newsletter, Feb. Vol. IV, pp. 53–55.

Kurian, A. (1999). Final Report of ICAR Adhoc Project "Evaluation and Selection of Medicinal Plants Suitable for a Coconut Based Farming System. Kerala Agri. Univ. Vellankkara, Thrissur.

Kurian, A. and Sankar, M.A. (2007). Medicinal plants - Horticulture science series Vol.-2, 1st ed., New India Pub. Agency, New Delhi 4 p.

Mader, T.L. and Brumm, M.C. (1987). Effect of feeding sarsasaponin in cattle and swine diets. J Anim Sci.; 65: 9–15.

Mishra, B. (1979). Bhava Prakash Nighantu (Indian Meterica Medica) commentary by K.C. Chunekar. 391 p.

Mishra, P. (1994). Biochemical investigation of Chlorophytum spp. (Safed musli). M.Sc. (Boichemistry) Thesis, (Published), RAU, Udaipur (Rajasthan). 41 p.

Negi, J. S.; Singh, P.; Joshi, G. P.; Rawat, M. S. and Bisht, V. K. (2010). Chemical constituents of Asparagus. Pharmacogn Rev.; 4(8): 215–220. doi: 10.4103/0973-7847.70921.

Newman, H.A.I.; Kummerow, F.A. and Scott, H.A. (1958). Dietary Saponin: A factor which may reduce liver and serum cholesterol level. Poultry Sci. 37: 42-46.

Nickavar B, Kamalinejad M, Haj-Yahya M, Shafaghi B. (2006). Comparison of the free radical scavenging activity of six Iranian Achillea species. Pharm Biol.; 44: 208–12.

Pandya, P. and Shambhudas (2007). Old and new herbal remedies. Shri Vedmata Gayatri Trust, Shantikunj, Haridwar (Uttrakhand). 38 p.

Patel, M.A., Patel, D.H., Patel, P.B. and Dalal, K.C. (1991). Studies related to domestication and collection of Chlorophytum spp. All India Coordinated Project on Medicinal and Aromatic Plants (ICAR). Workshop Report of GAU, Anand (Gujrat), Dec. 12–15. pp. 76–80.

Patwardhan B. (2003). AyuGenomics: Integration for customized medicine. Indian J Nat Prod.; 19:16–23.

Punia, M.S., Sharma, G.D., Verma, P.K. and Mer, B.R. (1980). Research on medicinal plants in Haryana, their retrospect and prospects. Proceedings of Regional Seminar on Medicinal plants, Manali, India, April 11-13, pp. 150–154.

Rao, M.R. (2005). New vistas in agroforestry. Agroforestry system, Pub. Springer Netherlands, pp.107–122.

Rastogi, R.P. and Mehrotra, B.N. (1999). Compemdium of Indian Medicinal Plants. Vol. 1-3. Central Drug Research Institute, Lucknow and National Institute of Science Communication, New Delhi.

Ravindran, P.K., Pillai, G.S. and Babu, K.N. (2004). Underutilized herbs and spices. In: Handbook of Herbs and Spices. Vol. 2. Eds. K.V. Peter. Woodhead Publishing Ltd., Cambridge.

Rocher, Y. (1965). Isolation of saponin from Ficaria ranunculeides. Chem. Abst. 64: 6418a.

Rochford, H. and Grover, R. (1961). The Rochford book of house plants. Asian Pub., New Delhi.

Sambasiva, K., Tripathi, H.C., Jawaharlal and Gupta, P.K. (1982). Influence of drugs on male sex behaviour and its pharmacological aspects: A Review. Indian Drugs. 19(4): 133–139.

Sanghi, D. and Sharma, N.J. (2001). Palynological studies of some medicinal plants. J. Phytological Res. 14(1): 83–90.

Saxena, V.K. and Chaurasia, S. (2001). A new isoflavone from the roots of Asparagus racemosus. Fitoterapia; 72:307–9.

Sekine, T.; Fukasawa, N.; Kashiwagi, Y.; Ruangrungsi, N. and Murakoshi, I. (1994). Structure of asparagamine A: A novel polycyclic alkaloid from Asparagus racemosus. Chem Pharm Bull.; 42:1360–2.

Sekine, T.; Fukasawa, N.; Murakoshi, I. and Ruangrungsi N. (1997). A 9,10-dihydrophenanthrene from Asparagus racemosus. Phytochemistry; 44: 763–4.

Sen, N.L. (2004). Importance and scope of medicinal and aromatic plants. In: Manual of winter school and commercial exploitation of underutilized MAPs. MPUA&T, Udaipur, pp. 1–6.

Seth, P., Sharma, M.K. and Simlot, M.M. (1991). Saponin in Chlorophytum spp. In: Paper presented in Diamond Jubilee Annual General Body Meeting of Society of Biological Chemists of India and Biotech. Institute of Chemical Biology, Calcutta, Dec. 27–30, Abstract No. 028.

Shah, C.S. (1982). Recent development of natural products. In: Cultivation and Utilization of Medicinal Plants. Eds. Atal, C.K. and Kapur, B.M. Regional Research Laboratory, Jammu. pp. 24-28.

Sharma, A.K. (2007). Medicinal and aromatic plants. Aavishkar Pub. & Distri., Jaipur (Rajasthan). pp. 62–79.

Sharma, P.V. (1978). Dravyaguna Vigan. Yug Nirman Yojna, Gayatri Tapobhumi, Mathura, pp. 559–566.

Shivarajan, V.V. and Balachandran, I. (1994). Ayurvedic Drugs and their Plant Sources. Oxford and IBH Publishing Co. Pvt. Ltd., New Delhi.

Shrivastava, J.L. and Pahapalkar, M.K. (1997). Cultivation Trials on Satavar (Asparagus racemosus) in Mandla district of Madhya Pradesh. Vaniki-Sandesh. 21(2): 24–33.

Singh, A. and Chauhan, H.S. (2003). Safed musli (Chlorophytum borivilianum): distribution, biodiversity and cultivation. J. Med. Arom. Plant Sci. 25: 712–719.

Singh, D. (1974). Udwij Aushadhdravya (Unani Dravyagunadharsh in Hindi). 587 p.

Singh, J. and Bangchi, G.D. (2002). Research strategies for development, standardization, quality control and good manufacturing practices (GMPs) for herbal drugs. Proc. First National Interactive Meet on Medicinal and Aromatic Plants, CIMAP, Lucknow. pp. 222–232.

Singh, J.; Srivastva, R.K.; Singh, A.K. and Kumar, S. (2000). Futuristic scenario in production and trade of major medicinal plants in India. J. Med. Arom. Plant Sci. 22–23 (4A-1A): 564–571.

Singh, P. and Singh S. (2007). Cultivation of safed musli: A medicinal cash crop. Intensive Agriculture. 45(1): 31–33.

Tarafder, C. R. (1983). Ethnogynecology in relation to plants use for abortion. J. Econ. Taxon. Bot. 4(2): 507–516.

Warrier, P.K., Nambiar, V.P.K. and Ramankutty, C. (Eds). (1995). Indian Medicinal Plants-a compendium of 500 species vol. 1-5. Orient Longman Ltd., Chennai.